Bioanalysis & Biosensors in AGRICULTURE SCIENCE

Bioanalysis & Biosensors in AGRICULTURE SCIENCE

Editor

P.B. Bansal

2006

Gene-Tech Books

New Delhi - 110 002

ISBN 81-89729-00-4

Published by : GENE-TECH BOOKS
4762-63/23, Ansari Road, Darya Ganj,
NEW DELHI - 110 002
Phone: 41562849
e-mail: genetechbooks@yahoo.co.in

Printed at : Tarun Offset Printers
Delhi

PRINTED IN INDIA

Preface

Biotechnology offers efficient and cost-effective means to produce an array of novel, value-added products and tools. It has the potential to increase food production, reduce the dependency of agriculture on chemicals, lower the cost of raw materials, and reduce the negative environmental impacts associated with traditional production methods. With improved technology and knowledge about agricultural organisms, processes, and ecosystems, opportunities will emerge to produce new and improved agricultural products in an environmentally sound manner.

The development of biosensors offers great promise for improving food processing, analysis, and safety assurance. The highly specific actions of biological molecules can be exploited for use in biosensors that can measure the concentration of specific components in complex mixtures. Enzymes, antibodies, and microbial cells can be immobilized on solid surfaces, and the specific reactions they mediate can be detected by various physical and chemical means. Future gains in agricultural productivity and sustainability depend heavily on the use of biotechnology to improve the health and well-being of agriculturally important plants and animals.

This book deals with the applications of modern biotechnology in the agriculture sector. It will be highly useful for students and practitioners from various fields, including agriculture, genetic engineering and environmental science.

Editor

Contents

1

Biosensors

A biosensor consists of two components. One is a bioreceptor and a trans-ducer. The bioreceptor is a biomolecule that recognizes the target analyte whereas the transducer converts the recognition event into a measurable signal. The uniqueness of a biosensor is that the two components are integrated into one single sensor.

This combination enables one to measure the target analyte without using reagents. For example, the glucose concentration in a blood sample can be measured directly by a biosensor by simply dipping the sensor in the sample. This is in contrast to the conventional assay in which many steps are used and each step may require a reagent to treat the sample. The simplicity and the speed of measurement is the main advantages of a biosensor.

Biosensor = bioreceptor + transducer.

Bioreceptor

Enzyme is a Bioreceptor. When we eat food such as a hamburger and French fries, they are broken down into small molecules in our body via many reaction steps (these breakdown reactions are called catabolism). These small molecules are then used to make building blocks of our body such as proteins (these synthesis reactions are called analbolism). Each of these catabolism and anabolism reactions (the combination is called metabolism) are catalyzed by a specific enzyme. Therefore, an enzyme is capable of recognizing a specific target molecule (Fig. 1). This biorecognition capability of the enzyme is used in biosensors. Other biorecognizing molecules (= bioreceptors) include antibodies, nucleic acids, and receptors.

Immobilisation of Bioreceptor

One major requirement for a biosensor is that the bioreceptor molecule has to be immobilized in the vicinity of the transducer. The immobilization is done either by physical entrapment or chemical attachment. Note that only minute quantities of bioreceptor molecules are needed, and they are used repeatedly for measurements.

Transducer: A transducer should be capable of converting the biorecognition event into a measurable signal (Fig. 2). Typically, this is done by measuring the change that occur in the bioreceptor reaction. For example, the enzyme glucose oxidase (used as a bioreceptor in a glucose biosensor) catalyzes the following reaction:

$$\text{Glucose} + O_2 \xrightarrow[\text{Oxidase}]{\text{Glucose}} \text{Gluconic acid} + H_2O_2$$

To measure the glucose concentration, three different transducers can be used:

1. An oxygen sensor that measures oxygen concentration
2. A pH sensor that measures the acid (gluconic acid) production
3. A peroxide sensor that measures H_2O_2 concentration.

Note that an oxygen sensor is a transducer that converts oxygen concentration into electrical current. A pH sensor is a transducer that converts pH change into voltage change. Similarly, a peroxidase sensor is a transducer that converts peroxidase concentration into an electrical current.

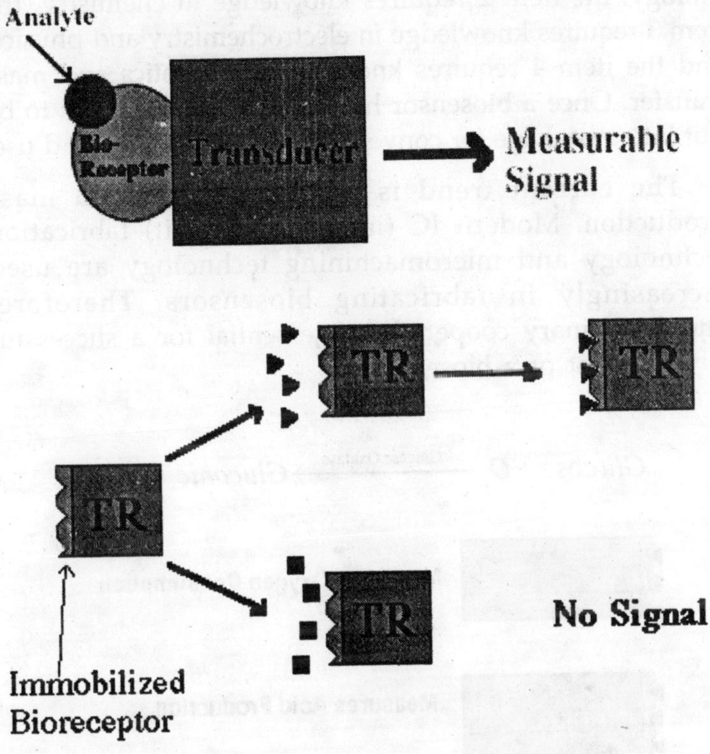

Figure 1. Specificity of biosensor

Considerations in Biosensor Development: Once a target analyte has been identified, the major tasks in developing a biosensor involves:

1. Selection of a suitable bioreceptor molecule
2. Selection of a suitable immobilization method

3. Selection of a suitable transducer
4. Designing of biosensor considering measurement range, linearity, and minimization of interference
5. Packaging of biosensor

The item 1 requires knowledge in biochemistry and biology, the item 2 requires knowledge in chemistry, the item 3 requires knowledge in electrochemistry and physics, and the item 4 requires knowledge in kinetics and mass transfer. Once a biosensor has been designed, it has to be put into a package for convenience manufacturing and use.

The current trend is miniaturization and mass production. Modern IC (integrated circuit) fabrication technology and micromachining technology are used increasingly in fabricating biosensors. Therefore, interdisciplinary cooperation is essential for a successful development of a biosensor.

$$Glucose + O_2 \xrightarrow{\text{Glucose Oxidase}} Gluconic + H_2O_2$$

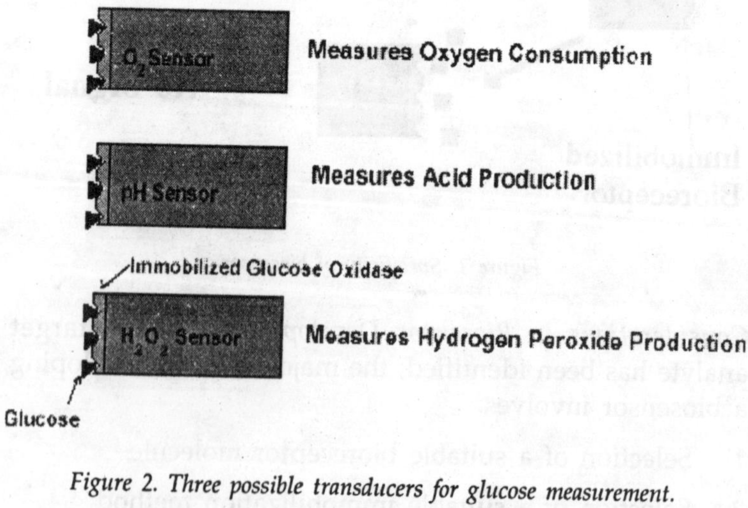

Figure 2. Three possible transducers for glucose measurement.

Requirements for Sensors: To be commercially successful, a biosensor has to meet the general requirements of commercial sensors. These are:

— Relevance of output signal to measurement environment
— Accuracy and repeatability
— Sensitivity and resolution
— Dynamic range
— Speed of response
— Insensitivity to temperature (or temperature compensation)
— Insensitive to electrical and other environmental interference
— Amenable to testing and calibration
— Reliability and self-checking capability
— Physical robustness
— Service requirements
— Capital cost
— Running costs and life
— Acceptability by user
— Product safety-sample host system must not be contaminated by sensor

Need for Biosensor

Diagnostic Market: Diagnostics represent a very large and well established market that is continually expanding. Particularly in the current climate of prevention rather than remedy, the need for detection at increasingly lower limits is increasing in many diverse areas. Estimates of the market size and future projections tend to be difficult and inaccurate.

Clinical Testing: However, undoubtedly clinical testing one of the biggest diagnostic markets. A study of the European market suggests a clinical testing products market in excess of 4000 million US$ in the 1990s (Biomedical Business International). In the US, the current biosensor market is already reported to be 12 million dollars, and future prospects vary from 100 to 10,000 million dollars by the turn of the century. This compares with a world market in 1985 of 1.5 billion dollars with an estimated growth rate of 9.5%, achieving a world market of 2 billion in 1990 and then expanding upwards and outwards.

Other Markets: Among the market shares, nearly 50% belongs to the medical arena (Technical Insights Inc.) with veterinary and agricultural applications amounting to a figure of half the size. Although the actual figures may differ between surveys, the relative proportions for different applications seem to be in agreement for the 1990s.

Table 1. Estimated markets in the 1990s

Market	Value (million dollars)
Medical and surgical	220
Veterinary and agricultural	105
Environmental and monitoring and safety	67
Industrial process monitoring	59

Figure 3. (Business Communications Co.) plots the US market projection into the next century and identifies a changing emphasis in applications with the agricultural

and environmental component becoming increasingly significant.

Applications of Biosensor

Health Care

Measurement of Metabolites: The initial impetus for advancing sensor technology came from health care area, where it is now generally recognized that measurements of blood gases, ions and metabolites are often essential and allow a better estimation of the metabolic state of a patient. In intensive care units for example, patients frequently show rapid variations in biochemical levels that require an urgent remedial action.

Also, in less severe patient handling, more successful treatment can be achieved by obtaining instant assays. At present, the list of the most commonly required instant analyses is not extensive. In practice, these assays are performed by analytical laboratories, where discrete samples are analyzed, frequently using the more traditional analytical techniques.

Market Potential: There is an increasing demand for inexpensive and reliable sensors to allow not only routine monitoring in the central or satellite laboratory, but also analysis with greater patient contact, such as in the hospital ward, emergency rooms, and operating rooms. Ultimately, patients themselves should be able to use biosensors in the monitoring and control of some treatable condition, such as diabetes.

It is probably true to say that the major biosensor market may be found where an immediate assay is required. If the cost of laboratory maintenance are counted with the direct analytical costs, then low-cost biosensor devices can be desirable in the whole spectrum of analytical applications from hospital to home.

Diabetes: The 'classic' and most widely explored example of closed-loop drug control is probably to be found in the development of an artificial pancreas. Diabetic patients have a relative or absolute lack of insulin, a polypeptide hormone produced by the beta-cells of the pancreas, which is essential to the metabolism of a number of carbon sources.

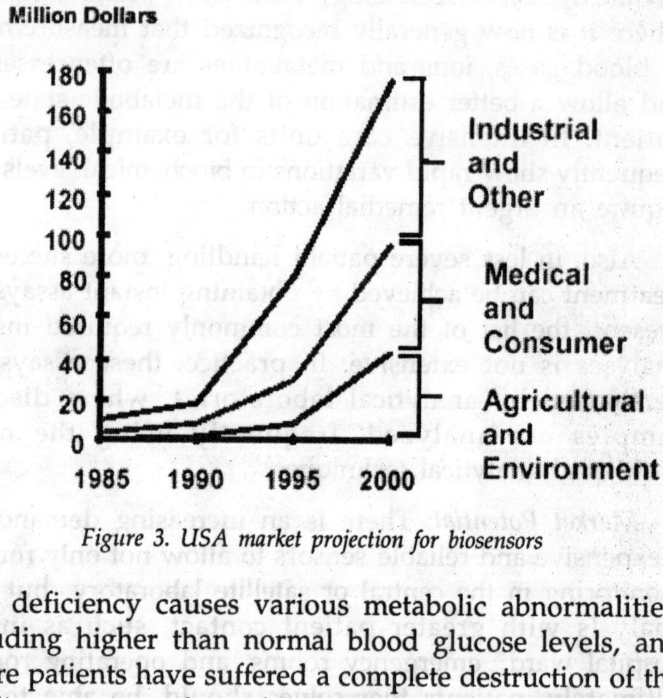

Figure 3. USA market projection for biosensors

This deficiency causes various metabolic abnormalities, including higher than normal blood glucose levels, and where patients have suffered a complete destruction of the insulin-secreting islets of Langerhans, then insulin must be supplied.

This has usually been achieved by subcutaneous injection, but fine control is difficult and hyperglycaemia cannot be totally avoided, or even hypoglycaemia sometimes induced, causing impaired consciousness and the serious long-term complications to tissue associated with this intermittent low glucose condition.

Insulin Therapy: Better methods for the treatment of insulin-dependent diabetes have been sought and infusion systems for continuous insulin delivery have been developed. However, regardless of the method of insulin therapy, its induction must be made in response to information on the current blood glucose levels in the patient.

Three schemes are possible (Fig. 3), the first two dependent on discrete manual glucose measurement and the third a 'closedloop' system, where insulin delivery is controlled by the output of a glucose sensor which is integrated with the insulin infuser. In the former case, glucose has been estimated on 'finger-prick' blood samples with a colorimetric test strip or more recently with an amperometric 'pen'-size biosensor device by the patient themselves. Obviously these diagnostic kits must be easily portable, very simple to use and require the minimum of expert interpretation.

However, even with the ability to monitor current glucose levels, intensive conventional insulin therapy requires multiple daily injections and is unable to anticipate future states between each application, where diet and exercise may require modification of the insulin dose. For example, it was shown that administration of glucose by subcutaneous injection, 60 min before a meal provides the best glucose/insulin management.

Artificial Pancreas: The introduction of a closed-loop system, where integrated glucose measurements provide feedback control on a preprogrammed insulin administration based on habitual requirement, would therefore relieve the patient of frequent assay requirements and perhaps more desirably frequent injections. Ultimately, the closed-loop system becomes an artificial pancreas, where the glycaemic control is achieved through an implantable glucose sensor. Obviously, the requirements

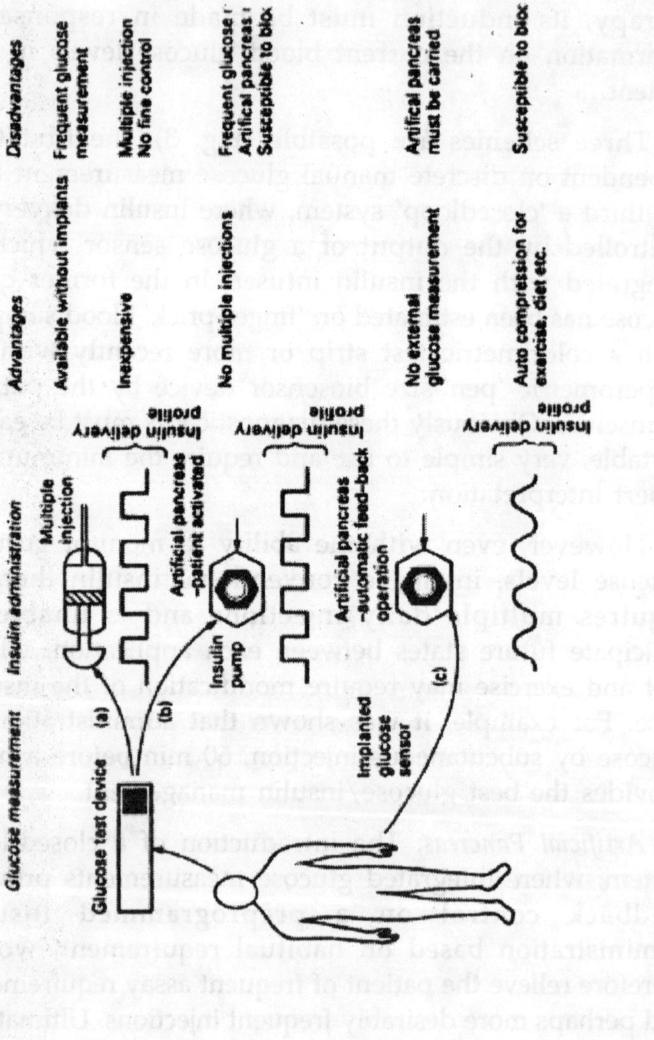

Figure 4. Schemes for insulin therapy

for this sensor are very different to those for the discrete measurement kits.

Seven requirements for an implantable glucose sensor.

— Linear in 0 - 20 mM range with 1 mM resolution

— Specific for glucose; not affected by changes in metabolite concentrations and ambient conditions

— Biocompatible

— Small—causes minimal tissue damage during insertion and there is better patient acceptability for a small device

— External calibration and < 10% drift in 24h

— Response time < 10 min

— Prolonged lifetime—at least several days, preferably weeks in use

Industrial Process Control

Bioreactor Control: Real-time monitoring of carbon sources, dissolved gases, etc., in fermentation processes (Fig. 5a) could lead to optimization of the procedure giving increased yields at decreased materials cost. While realtime monitoring with feedback control involving automated systems does exist, currently only a few common variables are measured on-line (e.g. pH, temperature, CO_2, O_2)) which are often only indirectly related with the process under control.

Three different methods of controlling a bioreactor are:

1. Off-line distant: central laboratory coarse control with significant time lapse
2. Off-line local: fine control with short time lapse
3. On-line: real-time monitoring and control

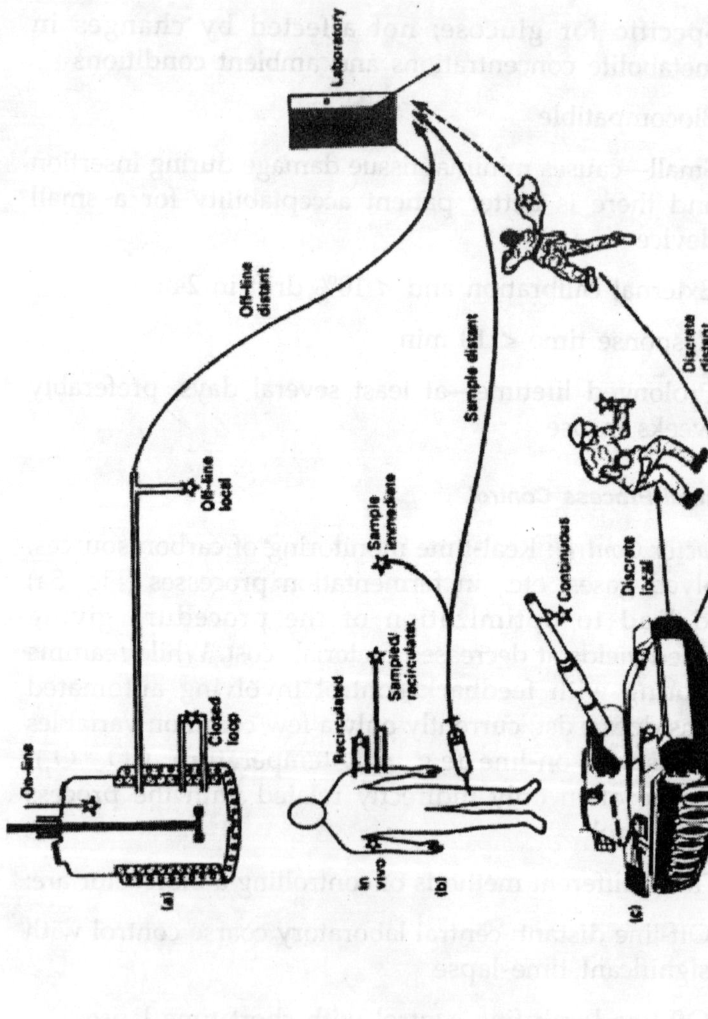

Figure 5. Comparison of sensing modes: (a) bioreactor; (b) clinical applications; © military or environmental monitoring.

On-Line Control: Method 3 is most desirable, which allows the process to follow an ideal pre-programmed fermentation profile to give maximum output. However, many problems exist with on-line measurements including in situ sterilization, sensor life-time, sensor fouling, etc. Some of the problems can be overcome if the sensor is situated so that the sample is run to waste, but this causes a volume loss, which can be particularly critical with small volume fermentations.

Off-Line Control: Although Method 3 may be the ultimate aim, considerable advantage can be gained in moving from Method 1 to Method 2 giving a rapid analysis and thus enabling finer control of the fermentation. The demands of the sensor are perhaps not as stringent in Method 2 as in Method 3.

Benefits of Control: The benefits which are achievable with process-control technology are considerable:

— Improved product quality; reduction in rejection rate following manufacture

— Increased product yield; process tuned in real time to maintain optimum conditions throughout and not just for limited periods

— Increased tolerance in quality variation of some raw materials. These variations can be compensated in the process-control management

— Reduced reliance on human 'seventh sense' to control process

— Improved plant performance-processing rate and line speed automated, so no unnecessary dead-time allocated to plant

— Optimized energy efficiency

The use of biosensors in industrial process in general could facilitate plant automation, cut analysis costs and improve quality control of the product.

Military Applications

Dip Stick Test: The requirement for rapid analysis can also be anticipated in military applications. The US army, for example, have looked at dipstick tests frequently two-step analyses taking up to 20 min to run. Such a time lapse is not always suited to battlefield diagnostics; the resulting consequences are suggested in Fig. 5(c). A particularly promising approach to this unknown hazard detection seems to be via acetylcholine receptor systems. It has been calculated that with this biorecognition system, a matrix of 13-20 proteins are required to give 95% certainty of all toxin detection.

Environmental Monitoring

Air and Water Monitoring: Another assay situation which may involve a considerable degree of the unknown is that of environmental monitoring. The primary measurement media here will be water or air, but the variety of target analytes is vast. At sites of potential pollution, such as in factory effluent, it would be desirable to install on-line real-time monitoring and alarm, targeted at specific analytes, but in many cases random or discrete monitoring of both target species or general hazardous compounds would be sufficient.

The possible analytes include biological oxygen demand (BOD) which provides a good indication of pollution, atmospheric acidity, and river water pH, detergent, herbicides, and fertilizers (organophosphates, nitrates, etc.). The survey of market potential has identified the increasing significance of this area and this is now substantiated by a strong interest from industry.

Potential applications for biosensors

— Clinical diagnosis and biomedicine

— Farm, garden and veterinary analysis

— Process control: fermentation control and analysis food and drink production and analysis

— Microbiology: bacterial and viral analysis

— Pharmaceutical and drug analysis

— Industrial effluent control

— Pollution control and monitoring o Mining, industrial and toxic gases

— Military applications

Tuning to Application: The potential for biosensor technology is enormous and is likely to revolutionize analysis and control of biological systems. It is possible therefore to identify very different analytical requirements and biosensor developments must be viewed under this constraint. It is often tempting to expect a single sensor targeted at a particular analyte, to be equally applicable to on-line closed-loop operation in a fermenter and pin-prick blood samples. In practice, however, the parallel development of several types of sensor, frequently employing very different measurement parameters is a more realistic.

Origin of Biosensor

Enzyme Electrode: The biosensor was first described by Clark and Lyons in 1962, when the term enzyme-electrode was adopted. In this first enzyme electrode, an oxido-reductase enzyme, glucose oxidase, was held next to a platinum electrode in a membrane sandwich (Fig. 6). The platinum anode polarized at + 0.6 V responded to the peroxide produced by the enzyme reaction with substrate.

The primary target substrate for this system was glucose:

$$Glucose + O_2 \xrightarrow{GlucoseOxidase} GluconicAcid + H_2O_2$$

and led to the development of the first glucose analyser for the measurement of glucose in whole blood. This Yellow Springs Instrument appeared on the market in 1974, and the same technique as employed here has been applied to many other oxygen mediated oxido-reductase enzyme systems.

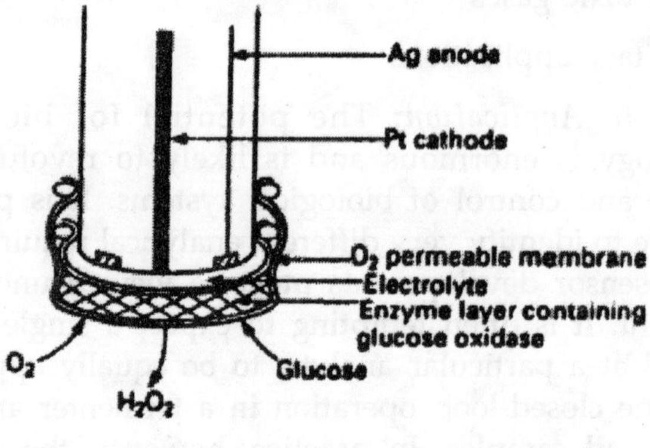

Figure 6. Clark enzyme electrode

Use of Membrane for Selectivity: A key development in the YSI sensor was the employment of membrane technology in order to eliminate interference by other electro-active substances. Polarized at +0.6V, the major interference to the peroxide measurement is ascorbic acid. Various combinations of membrane-enzyme sandwich have been developed, all satisfying the following criteria:

— the membrane between electrode and enzyme layer should allow the passage of H_20_2, but prevent the passage of ascorbate or other interferents

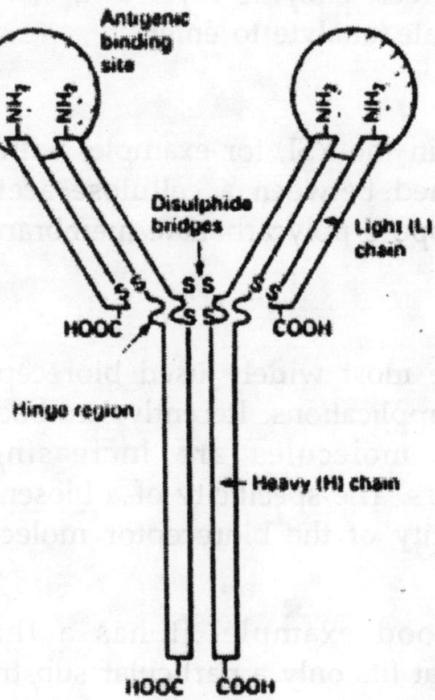

Figure 7b.

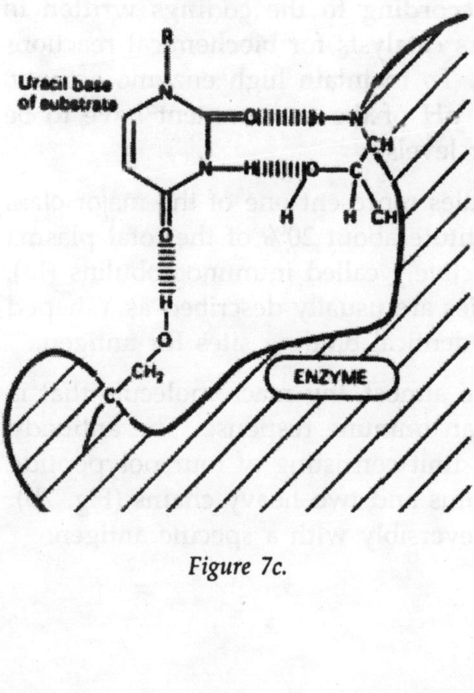

Figure 7c.

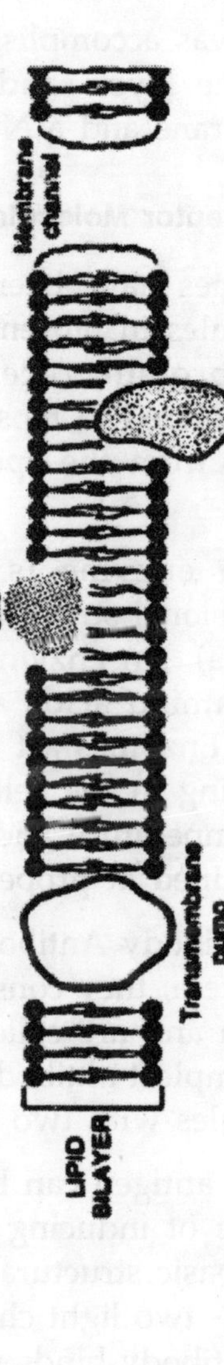

Figure 7a.

Figure 7. Bioreceptor molecules. (a) Enzyme, (b) Antibody, (c) Protein receptor

— the membrane between enzyme layer and sample should allow substrate/analyte to enter

— the enzyme layer

This was accomplished in the YSI, for example, with an enzyme layer sandwiched between a cellulose acetate membrane and a Nucleopore polycarbonate membrane.

Bioreceptor Molecules

Enzymes have been the most widely used bioreceptor molecules in biosensor applications. Recently, antibodies and protein receptor molecules are increasingly incorporated in biosensors. The specificity of a biosensor comes from the specificity of the bioreceptor molecule used.

An enzyme is a good example. It has a three dimensional structure that fits only a particular substrate (Fig. 7a). An enzyme is a protein synthesized in the cell from amino acids according to the codings written in DNA. Enzymes act as catalysts for biochemical reactions occurring in the cell. To maintain high enzyme activity, the temperature and pH of the environment have to be maintained at proper levels.

Antibody Antibodies represent one of the major class of protein; they constitute about 20% of the total plasma protein and are collectively called immunoglobulins (Ig). The simplest antibodies are usually described as Yshaped molecules with two identical binding sites for antigen.

An antigen can be almost any macromolecule that is capable of inducing an immune response. The antibody has a basic structural unit consisting of four polypeptide chains - two light chains and two heavy chains (Fig. 7b). The antibody binds reversibly with a specific antigen.

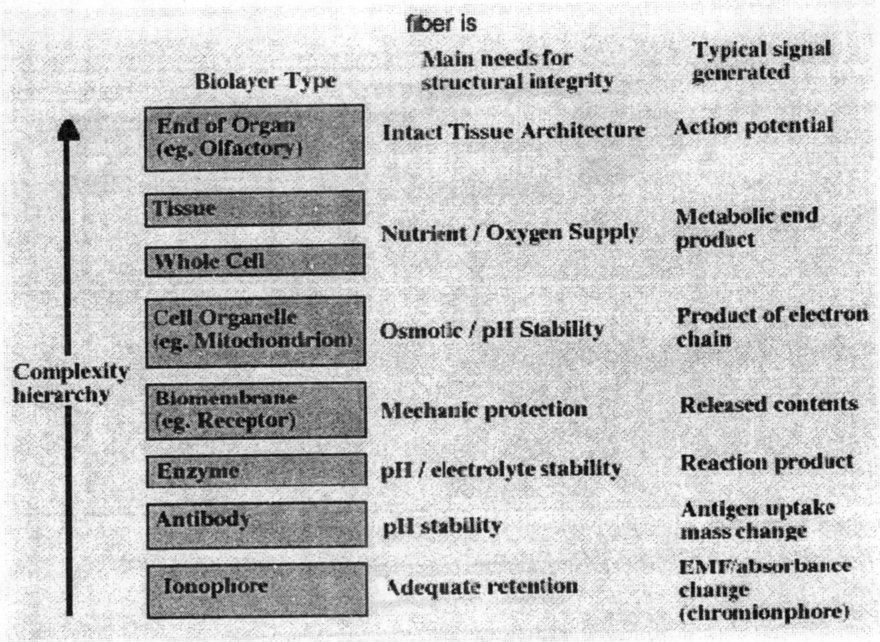

Figure 8. Possible bioreceptor molecules and molecular assemblies for biosensor applications; their requirements for structural integrity and signals generated.

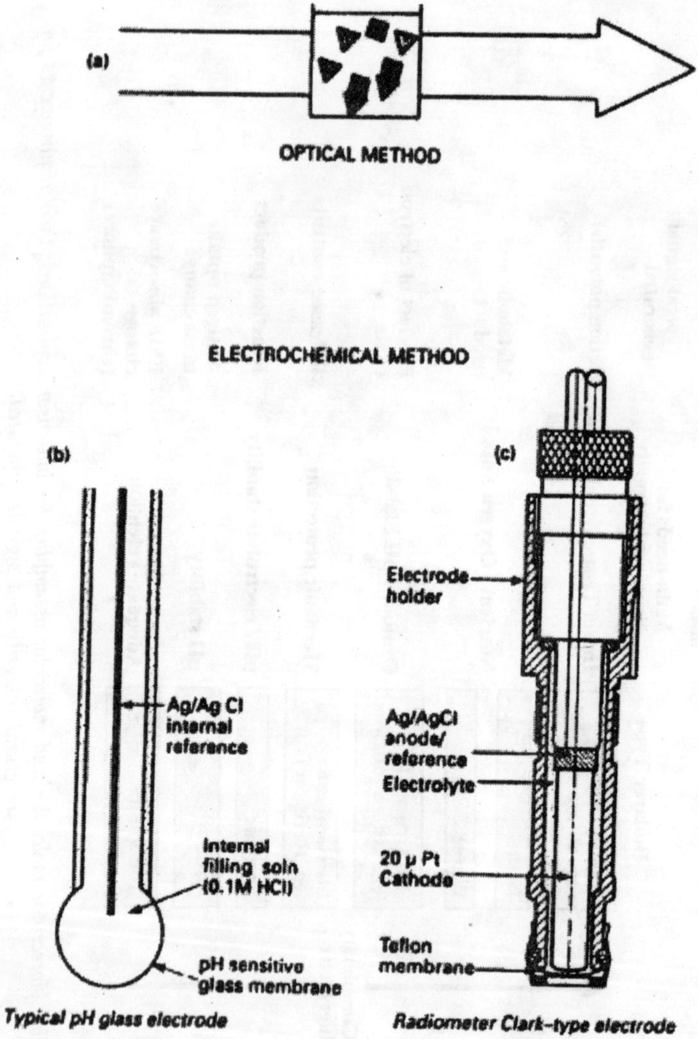

Figure 9. Three conventional transducers: (a) photometric; (b) potentiometric (based on pH sensor); (c) amperometric (based on Clark oxygen sensor).

Their purpose is to bind foreign substances - antigens - so as to remove them from the system.

Receptor Protein: Receptor proteins have specific affinity for hormones, antibodies, enzymes,and other biologically active compound. These proteins are mostly bound to membrane (Fig. 7c). There are hormone receptors, taste receptors, olfactory receptors for smelling, photoreceptors for eyes, etc. Receptor proteins are responsible for opening and closing of membrane channels for transport of specific metabolites.

Other Possibilities: In principle, any biomolecules and molecular assembly that have the capability of recognizing a target substrate (= the analyte) can be used as a bioreceptor. In fact, membrane slices or whole cells have been used in biosensors. Fig. 8 summarizes possible bioreceptors that can be utilized in a biosensor. Note that the bioreceptors require suitable environment for maintaining their structural integrity and biorecognition activity. These requirements are described in Fig. 8 along with the type of signal generated as a result of the biorecognition activity. The transducer in a biosensor has to be capable of measuring this signal.

Transducers Used

Conventional Transducers: Majority of biosensors existing today use three types of transducers for converting the action of the bioreceptor molecule into measurable signal. These are:

— amperometry based on H_2O_2 or O_2 measurement;

— potentiometry based on pH or pIon measurement;

— photometry utilizing optical fibers (Fig. 9).

Biorecognition reactions often generates chemical species that can be measured by electrochemical methods. In amperometry (Fig.8a), typically the reaction product is H_2O_2 (or the reactant is O_2) which can be measured by a pair of electrodes (Fig. 9a). When a suitable voltage is impressed on one of the electrode against a reference electrode (typically Ag/AgCl or Calomel), the target species (H_2O_2 or O_2) is reduced at the electrode and this generates electrical current (hence the name 'amperometry').

In potentiometry, a glass membrane or other membrane electrode is used for measuring the membrane potential (hence the name potentiometry) resulting from the difference in the concentrations of H^+ or other positive ions across the membrane. In photometry (Fig. 9c), the light from an indicator molecule is the measured signal. For this method to work, one of the reactants or products of the biorecognition reaction has to linked to colorimetric, fluorescent or luminescent indicator molecules. Usually, an optical used for guiding the light signals from the source to the detector.

Adaptation and exploitation of these three routes (photometric, potentiometric and amperometric) where user acceptability is already established, has been an obvious approach to the development of reagentless biosensor devices with a high specificity and selectivity.

Piezoelectric Transducers: The transducer of a biosensor is not restricted to the three described above. In principle, any variable which is affected by the biorecognition reaction can be used to generate the transduced signal. The piezoelectric materials and surface acoustic wave devices offers a surface which is sensitive to changes in mass. These transducers have been used where the biorecognition reaction causes a change in mass.

Conductimetry: Monitoring solution conductance was originally applied as a method of determining reaction rates. The technique involves the measurement of changes in conductance due to the migration of ions. Many enzyme-linked reactions result in a change in total ion concentration and this would imply that they are suitable for conductimetric biosensors.

Capacitance Measurement: When the biorecognition reaction causes a change in the dielectric constant of the medium in the vicinity of the bioreceptor, capacitance measurement method can be used as a transducer. Antigen-antibody reaction is a good example. Suppose antibody molecules are immobilized between two metal electrodes of known area. When antigen is added and binds with the antibody, the dielectric constant of the medium between the two electrodes is expected to change significantly. This change translates into a change in capacitance.

Thermometry: All chemical reactions are accompanied by the absorption (endothermic) or evolution (exothermic) of heat. Measurements of .H, the enthalpy of reaction at different temperatures allows one to calculate .S (entropy) and .G (Gibbs free energy) for a reaction and therefore collect basic thermodynamic data. The hydrolysis of ATP for example is exothermic:

$$ATP^{4-}+H_2O\sim ADP^{3-}+HPO^{4-}+H^+;\ \Delta H298 = -\ 22.2\ kJ\ (pH\ 7)$$

or the immunoreaction between anti-HSA and its antigen HSA yields –30.5 kJ/mol. For this latter reaction, the total increase in temperature for 1 ìmol of antibody is of the order of 10^{-5} K, but many enzyme-catalysed reactions have greater ΔH, and produce more easily measurable changes in temperature.

Enzyme Thermistor: For a biosensor device, the biorecognition compound must be immobilized on a temperature-sensing element capable of detecting very small temperature changes.

Initially, they immobilized glucose oxidase or penicillinase in a small column, so that temperature changes in the column effluent were monitored by thermistors to give an enzyme thermistor sensitive to glucose and penicillin, respectively. They have also applied the technique to other substrates and to immunoassay using an enzyme-labeled antigen.

FET as a Transducer: As advances are made in biosensors, there was a need for iniaturi-zation and mass production. Field effect transistors (FET) used extensively in semiconductor industry in memory chips and logic chips respond to change in electric field (in front of the 'gate' of FET). An FET is thus capable of detecting changes in ion concentration when the gate is expose to a solution that contains ions. Therefore, pH and ion concentration can be measured with an FET. The advantage of this transducer is that it can be incorporated directly to the electronic signal processing circuitry. In fact, pen-size FET based pH sensor is being marketed commercially.

Growth of Biosensor

Since the development of Clark's glucose sensor, many enzyme electrodes have been developed based on amperometry, potentiometry, and photometry. The term 'optode' is used for sensors utilizing optical fiber for light signal transmission. Note that the bioreceptors used are all enzymes except the antibody sensor. This represents the current state of the art in biosensor development - that bioreceptors other than enzymes are not explored extensively.

Common Products: The column 'Product detected' represent the type of transducer used. Note that common products (of biorecognition reactions) are used for measurement. For amperometry, the majority is H_2O_2 which is the common product for oxidoreductase enzymes.

For potentiometric biosensors, the majority is acid which can be detected by a pH sensor (CO_2 and NH_3 are indirectly detected by measuring the change in pH).

Biosensor Configurations: When bioreceptor molecules are combined with a suitable transducer, a biosensor is made. The bioreceptor molecules are immobilized in a suitable matrix to form a biolayer which is then placed in the immediate vicinity of a transducer.

The transducers ion-selective electrode and FET belong to potentiometric transducer category; the coated wire belong to amperometric sensor category; the surface plasma detector and the surface acoustic wave detector belong to piezo transducer category. The materials of constructions for the transducers are also given in the figure.

Discriminative Membranes: Membranes arc one of the essential component of a biosensor. They are used for

— preventing fouling;

— eliminating interference; and

— controlling the operating regime of the biosensor.

When a small molecule is the analyte, macromolecules such as proteins can be prevented to enter the active sensing zone by using a membrane that has small pore size. Note that proteins are notorious for causing fouling of the sensor. The transport of charged molecules can be controlled by using ion exchange type membranes. A combination of different discriminative membrane can also be used for blocking the passage of different interferents.

Sensitivity Requirements: The range and type of analytes are also varied and cannot be considered under a single umbrella. The particular application imposes a final concentration range requirement, but initially the concentration level that must be achieved can be estimated by the type of analyte of interest. Metabolites, for example, are commonly found at a level >10-6 mol/L, whereas hormones may be in the 10-10-10-5 mol/L range and levels as low as 10-20 mol/L would be desirable. For virus, 10-12 mol/L is desirable.

Immunoassay: An important stream of analytical developments, which has been widely applied, is the immunoassay techniques and the DNA probes. In immunoassay, the binding of antibody and antigen results in an increase in molecular mass and volume. Although current biosensor research is investigating the transduction of this phenomenon, the event is usually followed with a photometric, radioactive or even enzyme marker.

DNA Probe: In DNA probe assay, hybridization of strands of DNA antigen results in an increase in molecular mass and volume. The detection of this event is the same as those of immunoassay.

There are many major current reasons for replacing the radioisotopic labels with non-radioactive ones, but the direct use of photometric indicators have rarely provided the same degree of sensitivity, so that enzymes have to date frequently been proven to be the most promising form of labeling.

The principle of these label-linked assays is similar for both immunoassays and DNA probes. Both these techniques are heterogeneous assays - so that they are already developed along the lines of the biosensor concept.

Evolution of Biosensors: Biosensors can be classified into three generations according to the degree of integration of the separate components, i.e. the method of attachment of the biorecognition molecule (= bioreceptor) to the base indicator (transducer) element.

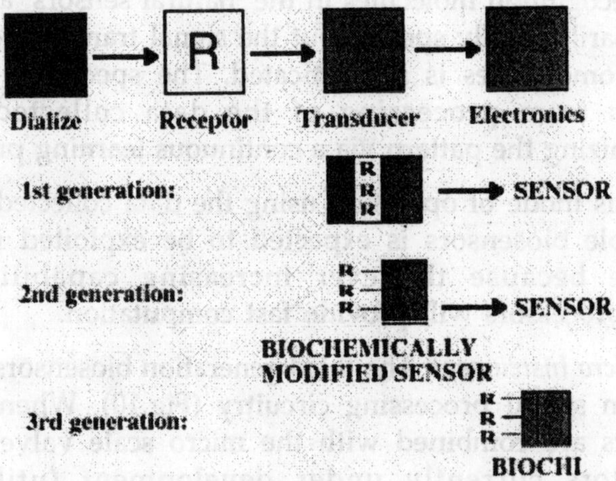

Figure 10. Three biosensor generations (R: Bioreceptor molecule)

In the first generation, the bioreceptor is retained in the vicinity of the base sensor behind a dialysis membrane, while in subsequent generations immobilization is achieved via cross-linking reagents or bifunctional reagents at a suitably modified transducer interface or by incorporation into a polymer matrix at the transduction surface. In the second generation, the individual components remain essentially distinct (e.g. control electronics-electrode-biomolecule), while in the third generation the bioreceptor molecule becomes an integral part of the base sensing element (Fig. 10).

While these definitions were probably intended for enzyme electrode systems, similar classifications appropriate to biosensors in general can be made. It is in the second and third generations of these families that the major development effort can now be seen.

Data Processing and Pattern Recognition. If we compare present biosensors with the natural ones (for example, the nose or the eye), they are very crude and simplistic.

The recognition molecules in the 'natural sensors' are not necessarily highly specific but the signal transduction via the biomolecules is sophisticated. The specificity often comes from processing of the data collected and recognizing the pattern via a continuous learning process.

This mode of operation using the data collected from multiple biosensors is expected to be exploited in the future because the ever increasing capability of microprocessors will provide fast computation.

Micro Instrument: The third generation biosensors have built-in signal processing circuitry (Fig.10). When such sensors are combined with the micro scale valves and actuators currently under development (utilizing micromachining technology), a whole analytical instrument can be built on a silicon wafer. Such an instrument can be mass produced and used in a variety of applications including homes, hospitals, automobiles, toxic dump sites, etc.

Molecular Electronics: The effort to continuously increase the density of electronic components to obtain ever smaller 'packages' will be limited eventually, not by the microlithographic technique employed but by the minimum size allowable for a transistor. Many biological molecules are able to synthesize complex self-organizing molecules with apparently just the required electronic properties. This suggests that the solution to this problem may be found in replacing silicon with biomolecular components. This idea has led to the proposition of many molecular electronic systems. In the past, materials and processing methods developed for microelectronic applications have been exploited in sensor developments.

Therefore, any future developments in molecular electronics are expected to be imported into biosensor technology. Multi-Disciplinary Nature. The arena of expertise required for biosensor development can be sustained by collaboration from many areas of academia and industry. The resulting output of this collaboration is likely in many cases to be a slow process, but is probably the only realistic route to successful future advances.

2

Biosensor Fundamentals

Glucose Biosensor

Use H_2O_2 Sensor for Transducer: Glucose sensor can be made in many different ways by using the enzyme glucose oxidase. This enzyme glucose oxidase catalizes the following reaction:

$$Glucose + O_2 \xrightarrow{GlucoseOxidase} GluconicAcid + H_2O_2$$

To measure the glucose concentration, three different methods can be used:

— Measure oxygen consumption by an oxygen sensor

— Measure acid (gluconic acid) production by a pH sensor

— Measure production of H_2O_2 by a peroxide sensor.

An oxygen sensor is a transducer that converts oxygen concentration into electrical current. A pH sensor is a transducer that converts pH change into voltage change. Similarly, a peroxidase sensor is a transducer that converts peroxidase concentration into an electrical current.

Structure of Biosensor: The structure of the biosensor to be made is shown in Figure 1. The base transducer consists of H_2O_2 sensor which is essentially the same as the oxygen sensor. The enzyme glucose oxidase is immobilized in front of the H_2O_2 sensor between two membranes.

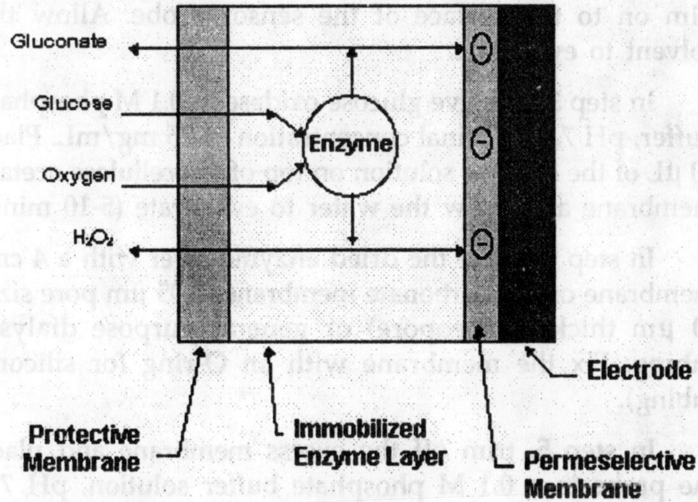

Figure 1. Overall structure of glucose biosensor.

The inner membrane is a permeselective membrane that allows passage of H_2O_2 where as the outer membrane separates the biosensor from measurement medium.

Enzyme Immobilization: The enzyme can be immobilized by:

— physical entrapment;

— cross-linking; and

— covalent attachment.

All three methods are to be used. The enzyme used is glucose oxidase in powder form.

Physical Adsorption

In step 1, prepare a mixture of 24 g of cyclohexanone, 24 g of acetone and 1 g of cellulose acetate (39.8% acetyl content, available from Aldrich). This is for casting the permeselective membrane.

In step 2, stir the mixture at room temperature until the cellulose acetate has dissolved and then cast a thin film on to the surface of the sensor probe. Allow the solvent to evaporate.

In step 3, dissolve glucose oxidase in 0.1 M phosphate buffer, pH 7.4 to a final concentration of 25 mg/mL. Place 20 μL of the enzyme solution on top of the cellulose acetate membrane and allow the water to evaporate (5-10 min).

In step 4, cover the dried enzyme layer with a 4 cm^2 membrane of polycarbonate membrane (0.05 μm pore size, 10 μm thick; Nucleopore) or general purpose dialysis tubing. Fix the membrane with an O-ring (or silicone tubing).

In step 5, trim off the excess membrane and place the probe in a 0.1 M phosphate buffer solution, pH 7.4 for 2hr before use.

Useful life of the sensor will be several months if the probe is stored in room temperature in phosphate buffer. The steps involved are shown in Fig.2.

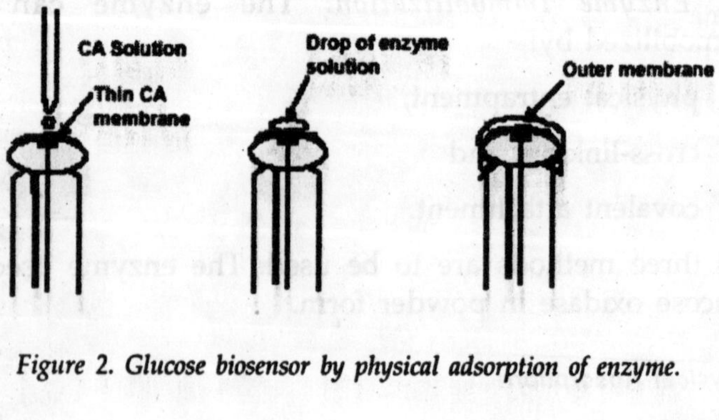

Figure 2. Glucose biosensor by physical adsorption of enzyme.

Crosslinking with Glutaraldehyde

Generally higher loading of active enzyme can be obtained if the adsorbed enzyme is cross-linked with glutaraldehyde. The method is the same as in physical adsorption method except one more step is added right after Step 3.

Step 3a. After the enzyme solution has dried, add 10 µL of a 1% solution of glutaraldehyde (Sigma Type I, supplied as a 25% solution which should be stored frozen and diluted in water immediately before use). Allow the glutaraldehyde solution to evaporate before proceeding to Step 4.

Crosslinking with BSA

Mutual crosslinking of the enzyme with another protein such as bovine serum albumin (BSA; Sigma, Fraction V powder). This procedure can lead to higher enzyme activity and greater stability. Step 3 of physical adsorption method is modified:

Step 3. Prepare the glucose oxidase solution as in Step 3 of Method 1. Also, prepare 50 mg/mL BSA solution in the same phosphate buffer. Mix 10 µL of each solutions and place the resulting 20 µL on the cellulose acetate membrane. After 1-2 min. add 10 uL of 2.5% glutaraldehyde solution.The liquid layer should harden rapidly. After 1-2 hr, go to Step 4 (Physical adsorption method).

Covalent Attachment to Membrane

Covalent attachment is most complicated but is useful in cases when the sensor is so small that the membrane must be fabricated directly on the sensing element. Covalent attachment gives more stable and reproducible enzyme activity.

Step 1: Dissolve 1.8 mg of cellulose acetate in a mixture of 20 mL of acetone and 3 mL of water.

Step 2: Cast 1 mL of this solution onto a clean dry glass plate using a spreader (Touzart) and allow the solvent to evaporate for 1 min at room temperature.

Step 3: Remove the membrane by immersing the glass plate in distilled water and floating it off. The resulting membrane is cut in to smaller pieces and stored at room temperature in water.

Step 4: Suspend four membranes (each 2.5 cm square) in 100 mL of 0.1 M sodium periodate for 20 min at room temperature.

Step 5: Wash the membranes in distilled water for 5 min then immerse them in 10 mL of a 10 mg/mL solution of BSA in 0.1 M borate buffer pH 9 for 2 h.

Step 6: Remove 9 mL of the BSA solution and add 4 mg of sodium cyanoborohydride (Aldrich). Incubate at room temperature for 2 h.

Step 7: Wash the membranes in distilled water for 5 min and then store in phosphate-buffered saline at room temperature.

Step 8: Recrystallize p-benzoquinone (Merck) from petroleum ether and prepare a solution of 15 mg/mL in ethanol.

Step 9: Add 100 μL of the freshly prepared p-benzoguinone to 0.5 mL of a 20 mg/mL solution of glucose oxidase in 0.1 M phosphate buffer pH 7.4 in a tube covered by aluminum foil.

Step 10: Incubate the mixture for 30 min at 37°C and then remove the excess pbenzoquinone by gel filtration through a Saphadex C-25 column (1 x 10 cm) equilibrated with 0.15 M sodium chloride and operating at a flow rate of 20 mL/h. Collect the pink-brown band that elutes in the void volume (2-3 mL).

Step 11: Suspend the BSA-cellulose acetate membranes in 2-3 mL of the activated glucose oxidase solution after adjusting the pH of the latter to 8 -9 with 0.25 mL of 1 M sodium carbonate. Incubate at room temperature for 38 h.

Step 12: Remove the membranes, wash them by stirring in 0. 15 S M potassium chloride solution for 24 h and then store them in phosphate-buffered saline pH 7.4 containing 1.5 mM sodium azide.

Design Variables

Immobilization Methods

Four methods are used for immobilizing enzyme for use in a biosensor:

1. adsorption;
2. entrapment;
3. covalent coupling; and
4. cross-linking.

These four methods are compared in Table 1. Among various methods, the crosslinking method is most frequently used because it has the advantage of the covalent bonding yet the cost is inexpensive.

Effects of immobilization

With immobilized enzymes the measured reaction rate depends not only on the substrate concentration and the kinetic constants K_M and V_{max} but also on so-called immobilization effects. These effects are due to the following alterations of the enzyme by the immobilization process.

Table 1. Comvarison of four enzume immobilization methods.

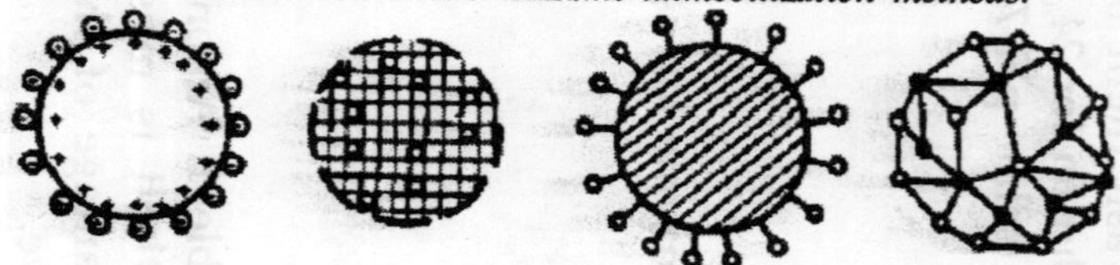

1 2 3 4

	1. Adsorption	2. Entrapment	3.Covalent coupling	4.Crosslinking
Matrix material	ion exchange resins, active charcoal, silica gel, clay, aluminum oxide, porous glass	alginate, carageenan, collagen, polyacrylamide, gelatine, silicon rubber, polyurethens	agarose, cellulose, PVC, ion exchange resins, porous glass	Crosslinking agents: glutaraldehyde, bisisocyanate, bisdiazobenzidine
Nature of bonding	reversible; changes in pH, ionic strength may detach the enzyme	physical entrapment	chemical bonding	entrapment; functionally inert proteins are often used together (BSA, gelatin)
Enzyme loading	low	low	high	high
Enzyme leakage	some	some	very low	low
Loss of activity	negligible	negligible	significant	small
Cost	inexpensive	inexpensive	expensive	inexpensive

— *Change in Conformation*: Conformational changes of the enzyme caused by immobilization usually decrease the affinity to the substrate (increase of K_M). Furthermore, a partial inactivation of all, or the complete inactivation of a part of the enzyme molecules may occur (decrease of V_{max}). These two cases of a conformation-induced drop of V_{max} may be distinguished by measuring the activity of the resolubilized enzyme or by titration of the active center with an irreversible inhibitor.

— *Change in Microenvironment*: Ionic, hydrophobic, or other interactions between the enzyme and the matrix (microenvironmental effects) may also result in changed K_M and V_{max} values These essentially reversible effects are caused by variations in the dissociation equilibria of charged groups of the active center.

— *Non-Uniform Distribution*: A non-uniform distribution of substrate and/or product between the enzyme matrix and the surrounding solution affects the measured (apparent) kinetic constants.

— *Reaction and Diffusion*: In biosensors the biocatalyst and the signal transducer are spatially combined, i.e., the enzyme reaction proceeds in a layer separated from the measuring solution. The substrates reach the membrane system of the biosensor by convective diffusion from the solution. The rate of this external transport process depends essentially on the degree of mixing.

In the multilayer system in front of the sensor the substrates and products are transferred by diffusion. Slow mass transfer to and within the enzyme matrix leads to different concentrations of the reaction partners in the measuring solution and in the matrix. Diffusion and the enzyme reaction do not proceed independently of one another; they are coupled in a complex manner.

Outer Membrane

Requirements

— The outer membrane has to be compatible with the medium into which it will be placed. Therefore, the requirement will be different depending on the nature of the measurement medium. For example, the outer membrane for biosensor used in liquid samples should be different from that intended for implantation application. For the latter application, bio-compatibility becomes an issue.

— The outer membrane should offer low diffusional resistance to analytes while the resistance should be high for macromolecules.

— For long-term continuous use applications, the fouling of the membrane must be minimal. The fouling causes an increase in the diffusional resistance of the analyte and thus the signal of the sensor changes as the fouling progresses. If microorganism grows on the surface of the outer membrane, the passage of oxygen to the enzyme layer is hindered which makes the sensor to behave erroneously.

Optimisation of Biosensor

Variations of the diffusion resistance of the semipermeable membrane are being used to optimize the sensor performance. A semipermeable membrane with a molecular cutoff of 10, 000 and a thickness of 10 μm only slightly influences the response time and sensitivity. In contrast, thicker membranes such as polyurethane or charged material, significantly increase the measuring time, but may also lead to an extension of the linear measuring range. Table 2. lists some of the commercially available membranes that can be used as the outer membrane.

Table 2. Available pre-cast membranes

	Collagen	Polycarbonate (Nucleopore)	Cellulose acetate
Characteristics	a hydroxylic natural protein	uniform pore size	slightly negative due to -COO-
Derivatizability	easy		easy
Temperature stability	O.K. at room T unstable at 37°C	stable	stable
Permeselectivity	exclude protein	exclude protein	exclude protein, retard transport of anionic species
Source	Sigma	Nucleopore	Amicon

Inner Membrane

Requirements: The inner membrane should be permeselective to target species only (for example H_2O_2 only for the current glucose sensor). Also, it should be as thin as possible and stable for long-term use. Some of the solution castable membranes and their characteristics are compared in Table 3.

Table 3. Solution castable membranes

	Cellulose acetate	Nafion	Polyurethane*
Characteristics		perfluorosulphonic acid ionomer, negatively charged	widely varying MW dep. on source
Solvent	acetone, cyclohexanone	low MW alcohols	dissolves in 98% tetrahydrofuran and 2% dimethylformamide
Coating method	dip coating, in-situ formation	dip coating as thin as	dip coating

		1000A (with 5% solution)	
Other characteristics		tend to adsorb proteins & cations, not useful as an outer membrane	biocompatible, retard glucose access to enzyme layer, passes O2 well

Enzyme Loading

Internal Diffusion is important: Usually in the operation of biosensors the flow conditions are adjusted to provide a mass transfer rate from the solution to the membrane system faster than that of in the enzyme layer. In the immobilized enzyme layer, reaction and diffusion occur simultaneously. Therefore, rigorous modeling is required to fully characterize the behavior of a biosensor. The key question in designing a biosensor is:

— how thick should the enzyme layer be? and

— how much enzyme has to be placed in the layer? Although rigorous modeling is required to fully characterize the behavior of biosensors, the design can be carried out by considering limiting cases.

Enzyme Loading Factor, f_E: The key variable for determining limiting cases is the enzyme loading factor defined by f_E:

$$f_E = \frac{(V_{max} / K_M)}{D_s / d^2} \tag{1}$$

where D_s and d are diffusivity of the substrate and the enzyme immobilization layer thickness. The enzyme loading factor is essentially the ratio of the reaction rate (for the case when $K_M >> S$; or first order reaction) to the diffusion rate (when $k = D_s/d$ and the mass flux J is expressed as k(dS/dx)).

Diffusion Control: $f_E > 25$: When the rate of reaction is much faster than the rate of diffusion, bottle neck of the transport process is the diffusion. In such a case, we say, the biosensor is operated in a diffusion-controlled regime. The condition is:

$$f_E > 25 \quad (2)$$

When this condition is met, the sensor signal depends on the diffusion process. This means the sensor output signal is linear with the analyte concentration, and is independent of the reaction rate of the enzyme layer.

Reaction Control: When the rate of reaction is much slower than the rate of diffusion, the bottle neck of the transport process is the reaction. In such a case, we say, the biosensor is operated in a reaction-controlled regime. The condition is:

$$f_E > 25 \quad (3)$$

When this condition is met, the sensor signal depends on the reaction rate and independent of the diffusion rate. This means that the sensor output signal depends on the reaction rate expression - i.e. the Michaelis-Menten equation. Therefore, the sensor output signal will be non-linear with the analyte concentration.

Advantages of diffusion control

There are many advantages of diffusioncontrolled biosensor. These include:

— Sensitivity is independent of enzyme content and activity

— Sensitivity is independent of inhibitors and pH variations.

— There is small temperature effect. Because the activation energy of diffusion is much less than that of reaction.

— Extended linear range. Extendible beyond $[S] = K_M$. For kinetics controlled case, linear range is achieved only for $[S] < K_M$

— Slower response time. The response time is determined by the diffusion time of the reaction product (sensed): d^2/D_p This is the major disadvantage of the diffusion controlled biosensor.

— Greater functional stability. Due to the excess of enzyme in the membrane, the sensor will have higher functional stability. With diffusion controlled sensor, 2000-10,000 measurement is possible, while kinetically controlled sensors permit only 200-500 measurement.

Apparent $K_{M:}$ When an enzyme is immobilized, the K_M value changes because of the changes in enzyme conformation and the microenvironment. Table 4.4. lists the K_M values for some of the immobilized enzyme systems.

Modeling of Biosensors

Modeling allows establishing the relationship between design parameters and the sensor signal output. In modeling, simplified and limiting cases are often useful to gain intuition and understanding of the sensing principle.

Amperometric Biosensor

Glucose Biosensor

Glucose biosensor using glucose oxidase is used as an example. The reaction and the transducer used are:

Reaction: Glucose + $O_2 \rightarrow$ Gluconic acid + H_2O_2

Transducer: polarographic H_2O_2 electrode

Model geometry

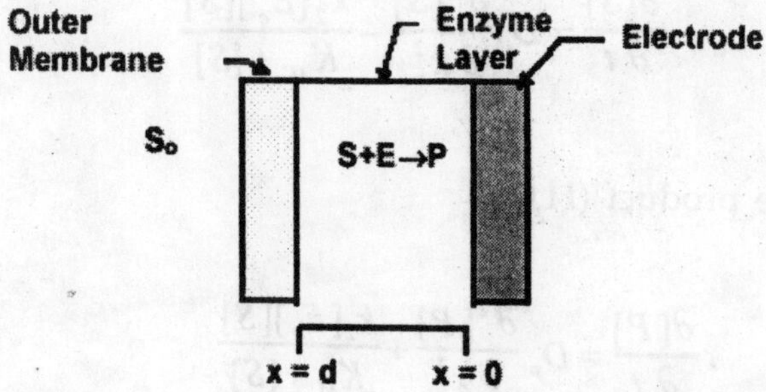

Assumptions: To write mass balance equations for the reaction and to set up boundary conditions, simplifying assumptions are necessary. Here, we will consider a very simple case for illustration purpose.

— The outer membrane offer no mass transfer resistance to the substrate (glucose) or the product (H_2O_2).

— No oxygen limitation in the enzyme layer.

— The substrate concentration at x=d is the same as that of bulk (no mass transfer resistance of the liquid film)

— The enzymes are homogeneously distributed in the immobilized layer.

— The inner membrane is extremely thin and offers no mass transfer resistance to the product or the substrate.

— The pores of the outer membrane is sufficiently large that it offers no mass transfer resistance to the substrate.

— One dimensional transport.

Differential mass balance

Unsteady state, differential mass balance is written for both the substrate glucose and the product (H_2O_2).

For the substrate (glucose):

$$\frac{\partial[S]}{\partial t}=D_s\frac{\partial^2[S]}{\partial x^2}-\frac{k_2[E_o][S]}{K_M+[S]} \tag{4}$$

For the product (H_2O_2):

$$\frac{\partial[P]}{\partial t}=D_p\frac{\partial^2[P]}{\partial x^2}+\frac{k_2[E_o][S]}{K_M+[S]} \tag{5}$$

Boundary conditions

To solve these equations, boundary conditions are necessary.

For the substrate:

@ t = 0, [S] = 0 (substrate diffusion has not started) (6a)

@ x = 0, $\frac{\partial [S]}{\partial x}=0$ (no transport of substrate) (6b)

@ x = d, [S] = $[S]_o$ (negligible liquid film resistance) (6c)

For the product:

@ t = 0, [P] = 0 (no product formation) (7a)

@ x = 0, [P] = 0 (rapid reaction of H_2O_2 by polarography) (7b)

@ x = d, [P] = 0 (well-mixing assumed) (7c)

When liquid film resistance is considered, replace Eqs. (6c) and (7c) by:

$$@\,x = d,\; D_S \frac{\partial [S]}{\partial x} = k_L([S]_\infty - [S]_o) \tag{6c'}$$

$$@\,x = d,\; D_P \frac{\partial [P]}{\partial x} = k_L([P]_o - 0) \tag{7c'}$$

Steady State Case When [S] << K_M

Since the solution process is complicated for the full unsteady state case, we will only solve the steady case when [S] << K_M

For the substrate (glucose):

$$D_S \frac{d^2[S]}{dx^2} = k[S] \tag{8}$$

where

$$k = \frac{k_2[E_o]}{K_M} \left(= \frac{V_{max}}{K_M} \right) \tag{9}$$

B.C. 1: @ x = 0, d[S]/dx = 0 (10a)

B.C. 2: @ x = d, [S] = $[S]_o$ (10b)

For the product (H_2O_2):

$$D_P \frac{d^2[P]}{dx^2} = -k[S] \tag{11}$$

B.C. 1: @ x = 0, [P] = 0 (due to rapid consumption)(12a)

B.C. 2: @ x = d, [P] = 0 (due to rapid mixing) (12b)

Solution of Eq. (5) with BCs (7a) and (7b):

$$[S]=[S]_o\left(\frac{\cosh(x\sqrt{k/D_S})}{\cosh(d\sqrt{k/D_S})}\right) \tag{13}$$

Substituting Eq. (13) in Eq. (11), and solving with BCs (12a) and (12b):

$$[P]=[S]_o\frac{D_S}{D_P}\left(\frac{x}{d}\left\{1-\frac{1}{\cosh(d\sqrt{k/D_S})}\right\}+\frac{1-\cosh(x\sqrt{k/D_S})}{\cosh(d\sqrt{k/D_S})}\right) \tag{14}$$

Sensor signal output will become:

$$I=nFAD_p\left(\frac{d[P]}{dx}\right)_{x=0}=nFAD_S\frac{S_o}{d}\left(1-\frac{1}{\cosh(d\sqrt{k/D_S})}\right) \tag{15}$$

Design Parameters on Sensor Performance

Solution in terms of f_E: The solution of Eq. (15) can be rewritten in terms of the enzyme loading factor f_E:

$$d\sqrt{k/D_S} = d\sqrt{\frac{V_{max}}{K_M D_s}}$$

$$= \sqrt{\frac{V_{max}\, d^2}{K_M\, D_S}} = \sqrt{f_E}$$

Therefore,

$$\boxed{I = nFAD_S \frac{S_o}{d}\left(1 - \frac{1}{\cosh(\sqrt{f_E})}\right)} \quad (16)$$

Sensor Output for Diffusion Control: When $f_E > 25$ (criterion for diffusion control), i.e.,

$$f_E = \frac{V_{max}\, d^2}{K_M\, D_S} = \frac{k_2[E_o]d^2}{K_M\, D_S} > 25$$

(17)

$$\boxed{I \cong nFAD_S \frac{S_o}{d}}$$

Key design parameter: Note that the key design parameter is f_E:

$$f_E = \frac{V_{max}\, d^2}{K_M\, D_S} = \frac{k_2[E_o]d^2}{K_M\, D_S}$$

To obtain diffusion control, how does one adjust V_{max}, d, and D_s?. Which variable is most effective?

Response Time: The response time of the sensor will be proportional to the diffusion time of the product τ_d:

$$\tau_d = \frac{d^2}{D_p} \tag{18}$$

Effect of enzyme loading factor, fE, on sensor output

Enzyme Reserve: The variation of the enzyme loading is a means of determining the minimum amount of enzyme required for maximum sensitivity. Furthermore, this test should reveal the magnitude of the enzyme reserve of diffusion controlled sensors.

Loading Test: Figure 4. shows the results of a loading test of GOD entrapped in a gelatin layer of 30 µm thickness between two dialysis membranes of 15 µm thickness each. The stationary currents for 0.14 mmol/L glucose (lower part of the linear measuring range) and for 5 mmol/L glucose (saturation) increase linearly with enzyme loading from 46 mU/cm2 to 1 U/cm^2. At higher GOD loading a saturation value is attained. To calculate the enzyme loading factor, f_E, the following values have been used:

<u>Parameters used</u>

d = 30 µm

K_M = 10 mM/L

D_S = 1.63x10^{-6} cm^2

A = 0.22 mm^2

Polarization: +0.6V

$$I = nFAD_s \frac{S_s}{d}\left(1 - \frac{1}{\cosh(\sqrt{f_E})}\right)$$

Parameters used

$d = 30\ \mu m$

$K_M = 10$ mM/L

$D_S = 1.63\text{x}10^{-6}\ cm^2$

$A = 0.22$ mm2

Polarization: +0.6V

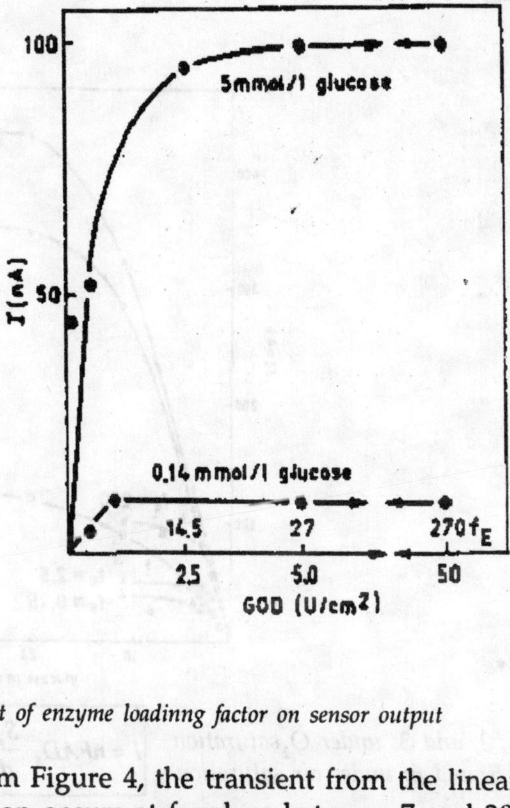

Figure 4. Effect of enzyme loadinng factor on sensor output

As is evident from Figure 4, the transient from the linear region to saturation occurs at f_E values between 7 and 20. This agrees with the theoretically predicted value and indicates that above $1U/cm^2$ the function of the GOD electrode is controlled by internal diffusion.

Concentration dependence of signal output

Linear Range: The linear measuring range of biosensors extends over 2-5 decades of concentration The lower detection limit of simple amperometric enzyme electrodes is about 100 nmol/L whereas potentiometric sensors may only be applied down to 100 µmol/L. This shows that the sensitivity is affected not only by the enzyme reaction but also by the transducer.

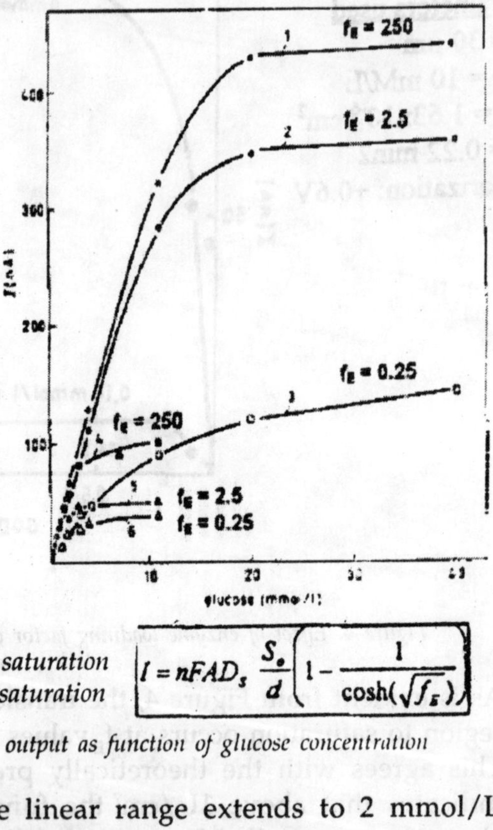

1, 2, and 3: under O_2 saturation
4,5, and 6: under air saturation

$$I = nFAD_s \frac{S_o}{d}\left(1 - \frac{1}{\cosh(\sqrt{f_E})}\right)$$

Figure 5. Sensor output as function of glucose concentration

Oxygen Effect: The linear range extends to 2 mmol/L glucose in the measuring cell. In this region, saturation of the measuring solution by oxygen increases the measuring signal by only 10%. At low glucose concentration the cosubstrate concentration (ca. 200 µmol/L at air saturation) influences the enzyme reaction only slightly. By contrast, in the saturation region above 2 mmol/L glucose the current rises by a factor of 4.5. At the same time the linear range is extended by oxygen saturation (Fig. 5).

Effect of pH on sensor output

With a high enzyme excess in the membrane, pH variations should have only a minor influence on the measuring process. Therefore the pH profiles in the linear measuring range and under diffusion control should be substantially less sharp than those of the respective enzyme in solution. The results obtained with a GOD-gelatin membrane (Fig.6) agree with this assumption. With 0.14 mmol/L glucose the curve is almost as flat as that of the H_2O_2 signal. On the other hand, with 10 mmol/L a pronounced maximum is found.

At this saturating concentration, the signal depends on the enzyme activity and therefore distinctly on pH. The pH optimum of immobilized GOD is about 0.9 pH units more alkaline than that of the soluble enzyme. Obviously the formation of gluconic acid within the enzyme membrane causes a local pH decrease, shifting the optimum to higher pH in the solution.

Effect of temperature on sensor output

The rate of enzyme reactions rises with temperature up to a certain optimum. Above that, the effect of thermal inactivation dominates over that of the increase of the collision frequency. Enzyme stabilization by immobilization is frequently reflected by an increase of the temperature optimum for substrate conversion. If kinetic and diffusion control are superimposed, the higher activation energy results in a predominant acceleration of the enzyme reaction with rising temperature. Thus, the slower enhancement of the diffusion rate makes mass transfer the limiting factor.

Therefore, the activation energy determined at lower temperatures is ascribed to the enzyme reaction, and that at higher temperatures to diffusion. Besides this, the

For $[S] << K_M$

$$I = nFAD_s \frac{S_o}{d}\left(1 - \frac{1}{\cosh(\sqrt{f_E})}\right)$$

where

$$f_E = \frac{V_{max}\, d^2}{K_M\, D_s} = \frac{k_2[E_o]d^2}{K_M\, D_s}$$

<u>Parameters used</u>

$d = 30\ \mu m$

$K_M = 10$ mM/L

$D_S = 1.63 \times 10^{-6}\ cm^2$

$A = 0.22$ mm2

Polarization: +0.6V

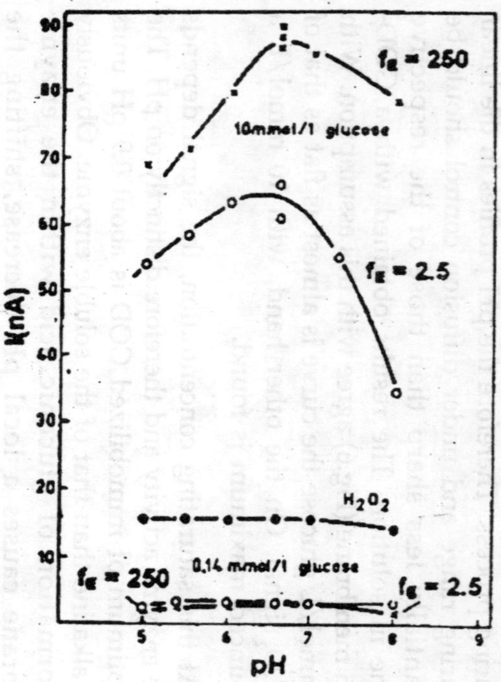

Figure 6. Effect of pH on sensor output

$$I = nFAD_s \frac{S_e}{d}\left(1 - \frac{1}{\cosh(\sqrt{f_E})}\right)$$

<u>Parameters used</u>

$d = 30\ \mu m$
$K_M = 10$ mM/L
$D_S = 1.63 \times 10^{-6}\ cm^2$
$A = 0.22$ mm2
Polarization: +0.6V

For H_2O_2 diffusion: $E_a = 33.5$ kJ/mol
For GOD reaction: $E_a = 25.5$ kJ/mol

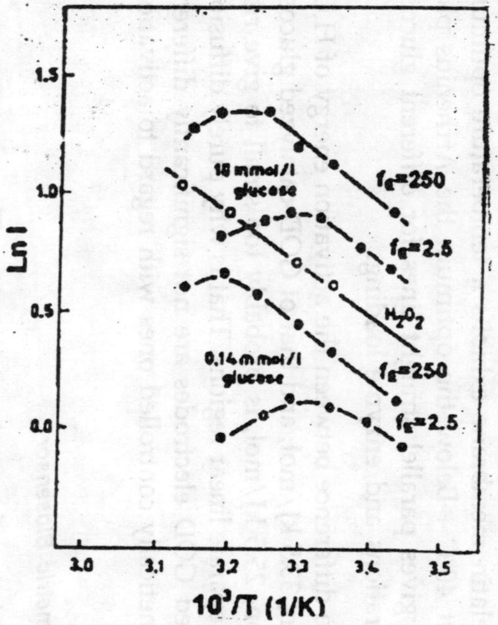

Figure 7. Effect of temperature on sensor output

temperature profile is affected by temperaturedependent conformational changes of the enzyme and decreasing solubility of the cosubstrate. The glucose sensor with the GOD-gelatin membrane exhibits a temperature optimum of about 40°C . Below the optimum the Arrhenius plot (Fig. 7) gives parallel straight lines for different glucose concentrations and enzyme loading.

The difference between the activation energy of H_2O_2 diffusion, 33.5 kJ/mol, and that of GOD-catalyzed glucose oxidation, 25.5 kJ/mol, is probably too small to give rise to two separate linear regions. That is why purely diffusion controlled GOD electrodes are not significantly different from kinetically controlled ones with regard to activation energy.

Potentiometric Biosensor

Characteristics of potentiometric sensor: The potentiometric sensor (the base sensor) has the characteristic that are different from amperometric sensors. These are:

— The measured species (such as H^+, NH_4^+, etc.) is not consumed.

— The sensor measures the activity (for dilute solutions, molar concentration can be used), of a specific ion (C_o) in reference to its internal standard (C_i).

— The output is in voltage (E_m).

— The ion specificity comes entirely from the membrane. For example, there is a membranes specific to H^+, one for NH_4^+, one for Ca^{++}, etc.

— The output of the base sensor is in voltage; independent of the sensor size; and proportional to natural log of C_o.

Examples of potentiometric Biosensor: Many biosensors have been developed using pH sensor as the base transducer.

Some of the examples and their performances are illustrated in the following.

Penicillin biosensor

$$\text{Penicillin} \xrightarrow{\text{Penicillinase}} \text{penicillic acid} + H^+$$

Response: 52 mV/decade over 5x10^{-2} to 10^{-4} M.

Glucose biosensor

$$Glucose \xrightarrow{Glucoseoxidase} gluconicacid + H_2O_2$$

Response: log linear response from 0.1 mM to 1 mM

Urea biosensor

$$Urea \xrightarrow{Urease} 2NH_4^+ + NCO_3^-$$

Response: log linear response from 5x10^{-5} to 5x10^{-3}

Configuration of biosensor

For potentiometric biosensors, the outer membrane is often not used.Often, the inner membranc is not used either. In such a case, the enzyme is immobilized directly on the surface of the potentiometric sensor.

Steady State Solutions for the Case When [S] << K_M

Equation and Bcs: The same equations (Eq. (8) through Eq. (12b)) are used to obtain the steady state solutions. The only change that has to be made is B.C.1 for the product. B.C. 1 has to be replaced by:

B.C. 1: @ x = 0, d[P]/dx = 0 (no transport; P is not consumed) The solutions are:

$$[S]=[S]_o\left(\frac{\cosh(x\sqrt{k/D_S})}{\cosh(d\sqrt{k/D_S})}\right) \tag{19}$$

$$[P]=[S]_o\frac{D_S}{D_P}\left(1-\frac{\cosh(x\sqrt{k/D_S})}{\cosh(d\sqrt{k/D_S})}\right) \tag{20}$$

Solution: Sensor signal output will be proportional to [P] at x = 0:

$$[P]_{x=0}=[S]_o\frac{D_S}{D_P}\left(1-\frac{1}{\cosh(\sqrt{f_E})}\right) \tag{21}$$

$$E_{out}=\frac{RT}{zF}\ln\frac{[P]_{x=0}}{[P]_i}=\frac{RT}{zF}\ln[P]_{x=0}+const \tag{22}$$

Sensor Output: For sensors under diffusion control (fE >25):

$$E_{out}=\frac{RT}{zF}\ln\frac{[P]_{x=0}}{[P]_i}=\frac{RT}{zF}\ln\left([S]_o\frac{D_S}{D_P}\right)+const \tag{23}$$

It has to be noted that under this condition, the sensor output Eout is proportional to the natural log the analyte concentration S, while the temperature dependence is linear:

$$E_{out}\propto \ln[S]_o \tag{24}$$

$$E_{out}\propto T \tag{25}$$

Effect of enzyme loading factor on sensor output

$$E_{out}=\frac{RT}{zF}\ln\frac{[P]_{x=0}}{[P]_i}=\frac{RT}{zF}\ln\left([S]_o\frac{D_S}{D_P}\left(1-\frac{1}{\cosh\sqrt{f_E}}\right)\right) \tag{26}$$

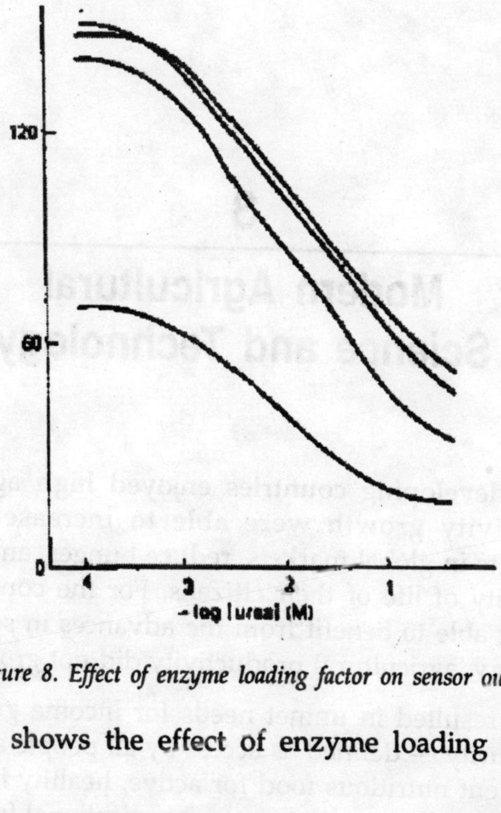

Figure 8. Effect of enzyme loading factor on sensor output

Figure 8 shows the effect of enzyme loading on sensor output.

3

Modern Agricultural Science and Technology

Today, developing countries enjoyed high agricultural productivity growth were able to increase incomes, participate in global markets, reduce hunger, and improve the quality of life of their citizens. For the countries that were not able to benefit from the advances in science and technology, agricultural productivity did not grow quickly.

This resulted in unmet needs for income growth and food security — defined as access by all people at all times to sufficient nutritious food for active, healthy lives. With supportive policy, regulatory, and institutional frameworks in place, science and technology can increase agricultural productivity and stimulate economic growth in all countries, thus reducing chronic hunger and offering more opportunities for participation in global markets.

Expanded global trade, investment, and economic integration could expand market opportunities for developing and developed economies alike. The potential benefits of international trade and technological progress are enormous. Integrated capital markets and the free flow of information create opportunities for growth and can have a significant impact on reducing poverty and hunger.

Agricultural Productivity

Technological advancement, broadly defined as any positive change in the way goods and services are produced, has been recognised by economists as a critical contributor to economic growth. Research is necessary to innovate, but product development and testing are needed before commercialisation or transfer of technology can occur.

Producers need good market and policy incentives to adopt new technologies, and the skills to use them effectively. These basic components of technology development and dissemination are the same in developed and developing countries. New technologies and innovative practices have been key factors in the economic development of high-income countries. Investments in agricultural research and development (R&D) by both the private and public sectors have resulted in a high level of productivity.

The production of more agricultural goods using fewer inputs frees resources to be invested in other parts of a country's economy, thus increasing affluence. Productivity increases occurred because of innovations in machinery, pesticides, fertilizers, information technologies, and plant breeding. While there has been a focus on production improvements for farmers, consumers also benefit from the increased production of basic commodities at low prices.

Innovations in food storage, processing, packaging, transportation, and increasing shelf life resulted in a wide variety of high-quality products being available year round. Recent breakthroughs in information technology and life sciences have expanded opportunities to increase production efficiency and to provide consumers with the safe, affordable, nutritious products they demand.

Consumers are increasingly concerned with the safety, variety, and nutritional value of food products. In addition, the public demands that agricultural production practices protect the environment and conserve natural resources. Some agricultural practices have had detrimental effects on human health and the environment.

Public research efforts have developed technologies and practices that have reduced these negative effects, and it is that set of technologies from which countries choose when trying to achieve their sustainable agricultural goals.

Green Revolution

The dramatic breakthrough in agricultural research in industrial countries, exemplified by yield gains and increases in agricultural productivity, took many years to reach some developing countries and bypassed others altogether. Before the 1960s, in developing countries, relatively little was invested in agricultural research, particularly for food crops. At that time, the Rockefeller and Ford Foundations helped establish an international agricultural research system to serve the research needs of developing countries.

The first efforts were in public research for rice, wheat, and maize. By the late 1960s, the development and spread of high-yielding varieties of these crops, combined with greater use of fertilisers and irrigation, led to notable increases in crop yields that greatly expanded the scope of the Green Revolution. This accomplishment reduced the incidence of famines, particularly in densely populated countries in Asia.

High-yielding varieties were developed by philanthropic or public research institutions and then given away or sold at low prices. Yield growth for various crops in developing regions has been substantial during the past three decades. For example, since 1965, wheat, rice, and maize yields in developing countries have more than doubled. doubled.

The contributions made by agricultural R&D to increasing food production, however, extend beyond yield increases alone. One of the major contributions of rice genetic improvement has been the development of varieties that produce yields similar to those of older rice varieties, but in shorter periods of time and with less loss of grain.

This has enabled double or even triple cropping in areas that previously produced only one or two crops per year. For other staple crops such as cassava, yield gains have been relatively modest. The net result of this R&D-driven technological transformation has been an increase in per capita food production in developing countries taken in the aggregate. From 1960 to 2000, for example, developing countries' population grew by around 125 percent, while the production of cereal in these countries tripled.

Over the same period, agricultural land in developing countries increased by only about 25 percent. Thus, increased yields per hectare, not the expansion of agricultural land, played the dominant role in expanding cereal production. In some regions, however, the expansion of agricultural land resulted in the loss of some ecological assets, but conservation efforts have been successful in slowing those losses in other areas.

India was able to increase conservation efforts and actually expand its forests and woodlands by 21 percent between 1963 and 1999. Economic studies have indicated that the rates of return to investment in agricultural R&D tend to be high in both developing and developed countries. Developing countries have made many impressive, scientifically based gains in food production over the past 40 years.

However, these successes have not been universal. Yield gains have been distributed unevenly among food crops that are important in developing countries. Much of the Green Revolution crop research resulted in advancements in wheat and rice production. Some maize/corn improvements were made that benefited part of Africa, but research did not focus on Africa's primary staple crops: yams, cassava, sorghum, and cowpeas.

Though yields for root crops like cassava have risen slowly since the 1960s, the rate of increase in yield has been much lower for these crops than it has been for cereals. Studies have shown that although small farmers lagged behind large farmers in adopting Green Revolution technologies, in many cases they eventually did benefit from the use of these innovations.

In both developing and developed countries, the farmers who can absorb the risks associated with trying new agricultural technologies due to their access to credit or larger holdings often adopt first. Their success serves as a model for those farmers who were initially uncertain about the new technology. Despite the many benefits to developing countries that resulted from the Green Revolution, there were some negative environmental impacts.

To effectively grow high-yielding crop varieties, fertilisers, pesticides, and water often were needed. Chemical residues were transported into waterways in tropical regions, and built up in soils in arid areas. Some chemicals leached into ground water. The use of synthetic pesticides had impacts on farm family health, and reduced the natural enemies of some targeted pests. These negative effects were experienced at the same time in developed countries, which led to research on technologies and practices to avoid the problems in the future.

Food Security and Income Growth

Many developing countries have a great need for increased productivity growth. Population growth rates in lower income countries are generally higher than in developed regions. If current trends continue, the world's population is expected to increase by 737 million people by 2011, and most of the growth will be in developing countries.

Unfortunately, crop yields are often substantially lower in these developing regions. Even though world food production has been increasing faster than population growth, many people are undernourished in less developed regions. In Sub-Saharan Africa, 43 percent of the population is chronically undernourished, consuming less than the minimum recommended nutritional requirements.

However, the greatest numbers of undernourished people live in Asia, which is the most highly populated region. With high population and low productivity levels, many low-income countries are not able to produce enough food domestically to meet basic nutrition needs. Nor do they have adequate income to import enough to eliminate these food gaps.

Agricultural productivity in developing countries must grow more rapidly than it has in the past decade, both to meet increasing demands for food and to raise rural and urban incomes-which, in turn, will lead to the possibility of increased agricultural trade and earning foreign exchange.

For example, since 1980, farm worker productivity rose by nearly 50 percent in Thailand, nearly doubled in China, and more than tripled in South Korea. These increases had significant effects on Asia's economies by stimulating growth, reducing poverty and malnutrition, and helping to keep food prices down. The development and adoption of new technologies will be necessary to increase both food supplies and access to food. Food-insecure countries with low incomes, reliance on local staple crops, and limited trade opportunities have benefited less than other developing countries from R&D-based advances in food production.

Many of the most significant advances in agricultural technology were made in developed countries where greater resources were devoted to agricultural R&D. Although food-insecure countries can be found in all major developing regions, they are particularly concentrated in Sub-Saharan Africa. In the lowest income countries, about one-third of food consumption comes from noncereal commodities such as cassava, for which there have been limited research investments and few technological breakthroughs.

In much of Sub-Saharan Africa, per capita food production has declined in the last two decades, a period in which public sector investment in agricultural R&D stagnated in this region. The World Bank estimates that 75 percent of the very poor, or nearly 1 billion people, live and work in rural areas and depend on agriculture for their livelihoods, either directly or indirectly.

Given that 90 percent of the food consumed in many developing countries is produced locally, production increases and product diversification could improve the health and well-being of the poor. Food security is the foundation for social security.

Potential Benefits

Developments in science and technology have contributed to better soil, nutrient, water, and pest management, and to more efficient methods of harvesting, storing, processing, and transporting farm products to market. Scientific breakthroughs have also occurred in our understanding of the complexity of sustainable agricultural systems, which has led to research into the development of sustainable crop management technologies and practices based on ecological principles.

Many factors influence technology adoption. Farmers choose from among alternative technologies and practices based on the biophysical characteristics of their environment, such as soil quality and access to water, as well as on social and economic characteristics such as land tenure, labor availability, income and wealth, profitability, and access to credit and information.

Many of the scientific and technological advances made in recent years potentially could be adapted to developing-country needs to increase productivity and environmental sustainability. Ultimately, the choice of appropriate technology will depend on the context in which it is used. It may not be the "newest" technology, but it could still fulfill the sustainable production goals of the country in which it is used. Many of these technology adaptations that are appropriate for smallholders will need to be provided by the public sector or public/private partnerships.

Technologies of Agricultural Production

Advances in soil and agronomic sciences have shown that application timing and method can be as important as input quantity for the effective use of fertilisers, pesticides, and irrigation water. Efficient input use results in fewer residues such as chemicals and salts accumulating in the environment. Knowledge of crop biological needs and resource conditions is often a critical input in crop management systems. There are many types of crop management systems, ranging from chemical-intensive practices to organic production systems.

The choice of appropriate practices to ensure a sustainable agricultural system depends on the characteristics of the environment in which the practices are used. Indigenous knowledge is important in designing an appropriate technology development plan, which needs to be in harmony with people, their societies and cultures.

Soil erosion

Soil erosion is not always visible and dramatic. In many areas, erosion by wind or water occurs slowly but steadily, and may not be recognised until damage is severe. In addition to the loss of productive soil, chemicals often adhere to soil particles and are transported by the erosion process to the environment. Tillage and land management systems have been developed to reduce soil disturbance, maintain optimal water-holding capacity, and increase soil nutrients and organic matter. Many of these technologies and practices can be adapted to meet the soil conservation needs of a developing country.

Water management

There are many demands for high-quality water supply for municipal, industrial, agricultural, and, increasingly, environmental uses. In 2000, worldwide freshwater use is estimated to have been about 70 percent for agricultural, 20 percent for industrial, and 10 percent for domestic use. However, Sub-Saharan Africa uses only 2 percent of its freshwater resources for irrigation.

The productivity of irrigated land is very high in both developed and developing countries. Water is the most common medium through which contaminants are transported to the environment. Whether through rainfall or irrigation, agricultural chemicals and nutrients can flow beyond the field or percolate to the water table.

Irrigation systems also can cause waterlogging, salinisation, and groundwater depletion. Therefore, efficient water management in all sectors is important for achieving a sustainable agricultural system. For individual farmers, the choice of irrigation methods is often limited by the water storage and delivery capabilities of the region, the quality of the land, water institutions, and investment requirements.

Irrigation technology innovations have contributed greatly to agricultural productivity, particularly in arid areas and for specialty crops. Technological improvements made in water storage and conveyance have reduced energy use and water losses. In some areas, small-scale irrigation projects have been very successful. On-farm low-volume systems such as drip irrigation, which was developed in the Middle East, have provided yield benefits in addition to per-hectare water savings.

Currently, equipment sensitivity and investment costs might make this technology inappropriate for some developingcountry applications, but fundamental lessons of water delivery efficiency and on-farm water management can be adapted to local needs. For example, irrigation scheduling based on a crop's evapotranspiration rate and actual weather conditions could conserve water while increasing yields.

Some low-cost irrigation technologies may be more beneficial in certain circumstances than those with higher investment or management requirements. The observations of an experienced farmer may be as effective as data from soil probes and meteorological stations in making an irrigation decision.

Pest management

The use of chemical pesticides in developed countries grew substantially after World War II. However, concerns about environmental contamination, ecosystem disruption, farm worker safety, and pest resistance led to substantial public research on alternative methods of pest management. If alternative pest control measures are not available, reductions in pesticide use may result in high production losses. The goal of integrated pest management (IPM) research is to design systems for controlling pest damage that are appropriate for the site while reducing reliance on chemical pesticides.

IPM programs often incorporate traditional practices, such as crop rotations, with sophisticated biological controls. Organic production systems do not use synthetic pesticides. Knowledge about pest biology and local environmental characteristics is important for the effective use of IPM.

The use of biologically based pesticides such as Bacillus thuringiensis (Bt) and the introduction (or reintroduction) of natural predators can be part of a sustainable pest management system, but these technologies require more knowledge and management skills than simple pesticide application. In addition, success of IPM programs requires that all farmers in the area work together. Community action has been an effective tool for implementing IPM plans in some developing countries.

Nutrient management

Soil's productive capacity depends on the nutrient content that is available to the crop. Natural amendments to soil have been used for centuries: ash, manure, crop residue, and seaweed. However, the most productive balance of nutrients often was not achieved. Even if the optimum amount of one nutrient is met, other nutrients may be in excess supply and leach into the environment. Improvements in chemical fertilizer technologies have enhanced farmers' ability to increase production in developed and developing countries alike.

Increased fertilizer use accounted for one-third of the growth in world cereal production in the 1970s and 1980s. Among developing regions, perhectare fertilizer consumption increased most rapidly in land-scarce areas (such as in Asia) and most slowly in Africa. Excess fertilizer components that were transported to the environment caused concern and led to research on better nutrient management practices.

Knowledge about soil chemistry and structure was used to design systems to sustain the productivity of the soil while reducing nutrient losses to the environment. Technologies to test soil and plant tissue nutrient content have been improved to give farmers timely information that can be used in making decisions. These technologies can greatly enhance the efficiency of the use of manure, which is an important source of nutrients in many developing countries.

Biological knowledge is used to tailor nutrient applications to plant growth needs, in terms of both timing and quantity. Application technologies have been improved to deliver the nutrients close to the root zone, which increases the amount available for uptake by the plant while reducing losses of nitrogen to air and water resources. Much of what has been learned about the chemical and biological aspects of nutrient management can be used to design systems that are in harmony with a country's sustainable agriculture goals.

Improvements of Crop

Increasing the yield potential and desirable traits in crops has long been a goal of agricultural science. Humans have been altering the genetics of their food supply since plants and animals were first domesticated thousands of years ago. About half of all recent gains in crop yields are attributable to genetic improvements. Innovations in plant breeding made in the public sector and international agricultural research centers after World War II produced the Green Revolution in many parts of the world.

Plant breeders have succeeded in developing crop varieties with high yields that will produce under particular pest pressures or environmental stresses. To obtain these benefits, however, investments in complementary crop management technologies such as irrigation or fertilizer use may be necessary. In addition, there is usually a gap-and it may be wide-between yields obtained in a laboratory or a controlled field trial and those actually experienced by farmers in their environment.

Many innovations have to be adapted through further research, experimentation, and farmer involvement. Even with these efforts, there may be a need for major investments in complementary crop management technologies before yield or quality goals are reached. In addition, any new variety needs to be assessed with respect to its potential impact on the biological environment, such as its contribution to pest resistance, unwanted gene flow, or loss of biodiversity.

At the end of the 20th century, breakthroughs in molecular biology led to modern biotechnology and the development by the private sector of crops that are diseaseand pest-resistant or herbicide-tolerant. Genetic engineering can increase productivity and achieve higher levels of stability and sustainability. Current farm-level biotechnology research is focused on developing crops that will tolerate a wider range of drought, acidity, salinity, heat, and flooding. These crops could contribute to productivity increases in resource-poor countries.

For example, with the help of genetic engineering, scientists are developing a virus-resistant sweet potato and pest-resistant variety of cassava. In South Africa, where 7 of every 10 cotton farmers have switched to biotechnology-derived varieties, farmers report that their production costs have decreased, and they use fewer pesticides. Also, the resulting reduction in tillage allows the soil to retain more water. Insect-resistant maize is being grown successfully by some small farmers as part of a pilot project supported by a biotechnology company.

Biotechnology tools also can be used for much more than just the production of bioengineered plants or animals. Tissue culture is the biotechnology tool used most frequently in developing countries. Many improvements in staple crops important to African people have been made recently with tissue culture. Molecular marker-aided selection methods can greatly speed the traditional plant breeding process, which can help crops respond more rapidly to pest pressures or environmental changes.

Desired traits can be identified early in a plant's development rather than having to wait until maturity to observe the trait. This ability is particularly important for those plants and animals that take years to reach maturity. Monoclonal antibody technology uses immune system cells to make proteins called antibodies, which can be used as a diagnostic tool to locate substances that occur in minuscule amounts.

Another technology with similar uses for detecting and monitoring involves biosensors, which are composed of a biological component linked to a tiny transducer. Biotechnology can be used to create bioengineered plants that can be used as manufacturing "facilities" for pharmaceutical compounds such as therapeutic proteins and vaccines. For example, researchers have developed a vaccine for hepatitis B that is produced by a banana for a fraction of the cost of a traditional vaccine.

Crop plant production of these products may lower costs and increase supply compared to current pharmaceutical production. For developing countries, vaccine-producing plants might be easier to grow locally and make available to rural populations than current vaccines.

Precision agriculture

Precision agriculture is typically characterised as a suite of information technologies used to monitor and manage sub-field spatial variability. Farmers use satellite technology, computers, and robotics to manage the use of pesticides, fertilizers, and water more efficiently by tailoring input amounts to the specific characteristics of the site. The benefits of precision agriculture technologies are greatest when field or farm conditions vary widely and the uniform applications of inputs will result in production inefficiency.

Each location is tested and a site-specific management plan is designed for individual conditions. Soil testing and field mapping can be used to identify places in a field where additional nutrient use will increase yield, or where input use can be reduced while maintaining yield. Variable-rate application of seeds, fertilizers, pesticides, and irrigation water has the potential to enhance producers' profits by reducing input costs. It may also reduce the risk to the environment from agricultural production by tailoring input use and application more closely to ideal plant growth and management needs.

In addition, by improving the efficiency of input use, precision farming has the potential to reduce the transport of agricultural chemicals through surface runoff, subsurface drainage, and leaching. Because the investment cost is high relative to the value of information received, current precision farming technologies are not likely to be appropriate for use by farmers with small holdings in developing countries. The systems could, however, be of great value for technology development planners, especially in assessing the productive capacity of natural resources and the appropriate suite of technologies and practices to sustainably increase production.

Animals/livestock

Selective breeding has been used worldwide to increase production of those animals that are most productive for the environment in which they live. For several decades, a more reliable technology has been used where the sperm and eggs are taken from bulls and cows with genetically preferred traits. These cells are united in the laboratory and cultured before being implanted in surrogate cows.

The results of this breeding method are more reliable in getting enhanced traits, and the quantity of desired offspring can be increased. Animal health research has been an important factor in increasing productivity and product quality. Particularly in an era of global mobility, the rapid and accurate diagnosis of disease can slow the spread of infection. Quarantines and embargoes are most effective before a problem becomes widespread.

Biotechnology-based diagnostic tests are more sensitive and easier to transport than older diagnostic methods. Diagnosing diseases such as brucellosis, pseudorabies, avian leucosis, or foot-and-mouth disease sooner and with greater accuracy means that appropriate therapy can be started sooner, thus decreasing the spread of the disease. Low-cost diagnostic technologies would be a valuable tool in the more isolated rural areas in developing countries where livestock health is critical for food security.

In developed countries has the benefits of integrating animal and crop production systems. By growing feed crops for their own animals, producers control the quality of the feed and may save on the purchase of inputs. In addition, the livestock waste can be used to increase soil quality. These integrated systems have been used throughout the developing world, but application of new scientific findings can increase productivity and contribute to higher environmental quality. New technologies for nutrient testing and manure spreading reduce runoff and leaching of animal wastes and increase fertilizer benefits to the crop.

Forestry and biomass

Although forests are harvested to produce building supplies and paper products, there has been an increasing appreciation of standing forests as a valuable ecosystem that can provide biodiversity, wildlife habitat, recreational opportunities, and a source of carbon sequestration. Lowimpact logging practices are being developed to replace clear-cutting practices and to preserve the integrity of the forest ecosystem while harvesting wood products.

Conversion of forests for agriculture accounted for two-thirds of the world's deforestation during the last 20 years. The most successful reforestation and agroforestry projects have resulted from longterm planning by rural communities that were committed to improving the local natural resource base.

In many developing countries, trees and other woody vegetation are the primary source of fuel in rural communities. The need to gather wood in resource-poor areas often causes women and children to travel long distances carrying heavy loads. The production of accessible and sustainable sources of fuel could free household resources for other uses.

There is increasing interest in developing sources of biomass to substitute for fossil fuels. Bio-feedstocks can be relatively clean-burning and have less waste than petroleum-based fuels. In addition, extraction/harvest, when properly managed, has a low environmental impact.

The technologies developed to improve the economic feasibility of biomassderived energy may be useful for developing countries.

Aquaculture

One of the fastest growing segments of the world's food production is aquaculture. It represents an alternative to the wild harvest of some fish species that are threatened by pollution and overfishing, and provides an excellent source of protein. Aquaculture is the production of aquatic animals and plants under controlled conditions for all or part of their life cycle. Sometimes the level of control is minimal, while in other cases the environment is designed to mimic a closed ecosystem.

For example, in shallow coastal waters, a frame can be placed to catch natural oyster spawn and the shellfish mature on the frame where they are easily harvested. In some areas, coastal oilrigs have inadvertently become shellfish nurseries. The other extreme in aquaculture technology is the use of "farms" where fish are grown in tanks of constantly tested, filtered water. The feed is developed to meet the nutritional needs of the fish, and the animals are monitored for diseases that affect fish productivity and for organisms that might pose health hazards to consumers.

A variation on this technology system for the intensive production of fish is called aquaponics, which combines the concentrated production of a vegetable or fruit crop as part of the recirculating system. Nitrogen waste from fish metabolites provides nutrients to the crop. By removing these wastes, the vegetation filters and cleans the water, which promotes faster fish growth.

Uses of Technologies

The main uses of technologies are Marketing, Processing,

and Transportation. In developed countries, the choices of foods that are readily available to consumers have expanded greatly in the past two decades. Innovations in storage, transportation, processing, and marketing have made the increase in affordable products possible. In these countries, much of the research on postharvest technologies has been done in the private sector.

Many of these innovations can be adapted for developing-country needs through collaboration with a range of institutions from developed and developing countries. Increased production on the farm will not yield sufficient benefits if the products are not delivered in an acceptable form and a timely way to an end-user. Major innovations in transportation have reduced the costs of long-distance trade and have increased opportunities for expanding markets.

These reductions in transportation cost will lower the price of the product and make it more available both geographically and economically. The objective of product transportation is to get products to market while maintaining quality and reducing handling and time in transit. The use of large, standardised containers that can be transported by truck, train, and ship without repacking has significantly lowered transportation costs in developed countries. For developing countries, containerisation allows ports to greatly increase shipping capacity.

The shelf life of fresh fruits and vegetables and their durability for transport have been increased, although in the past these traits often came at the expense of taste. Edible food films have been developed to reduce spoilage and dehydration of fresh fruits and vegetables. Recent genetic research is focused on targeting the desired traits that consumers demand.

In developing countries, affordable, small-scale technologies for preserving vitamin-rich fruits and vegetables can help combat micronutrient deficiencies by ensuring an adequate diet throughout the year. Technologies are also needed to accumulate, treat, and deliver perishable commodities such as milk that are produced on many geographically dispersed small holdings.

Foodborne illnesses are caused primarily by micro-organisms such as bacteria, viruses, molds, and parasites. Food safety hazards can come from unclean water, lack of refrigeration, and unsanitary conditions for food transport, storage, marketing, and preparation. Food safety can be increased by the use of new technologies. The use of biosensors in processing plants reduces contaminants and enhances quality. Also, product quality characteristics can be identified with the use of sensory panels.

Irradiation will reduce food-borne pathogens and may be an effective method to use for fresh produce, especially when pre-harvest contamination from the use of manure fertilizers may occur. Computer and communication technologies have improved quality control for production and the ability to market agricultural products efficiently. Food safety monitoring technologies and sanitary practices have greatly reduced microbial contamination. Rapid testing for mycotoxins, pesticides, and other environmental contaminants is extremely important in meeting international quality standards.

Production wastes are being reduced or recycled more frequently than in the past. Food-processing technologies have been used to transform raw agricultural commodities to meet consumer demands. Convenience is one characteristic that consumers have requested, along with enhanced flavor and nutritional content. Raw materials are being produced from traditional plant breeding and biotechnology methods to have higher contents of desired processing traits such as oils or starch, and lower amounts of other traits such as allergens.

In developed countries, several staple foods, such as bread and milk, are routinely fortified with vitamins. These additions have drastically reduced the incidence of rickets, scurvy, goiter, and other afflictions caused by nutritional deficiencies. Research is active in the area of functional foods that contain biologically active components that impart health benefits, which will eliminate the need to add the components later. This breakthrough would be particularly valuable for rural populations who consume food locally rather than purchase processed food.

Many have described the beginning of the 21st century as the Information Age. Precision farming and biotechnology resulted from the increased ability to analyse information. Innovations in computing capabilities and low-cost access to computers have dramatically enhanced the ability to store and analyse data. In addition, today's communication networks allow the rapid exchange of information. Firms can assess consumer demands worldwide, farmers can produce value-added crops for specific markets, and scientists can collaborate with researchers around the world to gather and analyse data.

Developments in multiple scientific disciplines have led to exciting discoveries, and to the origin of several new fields: bioremediation, nanotechnology, genomics, and bioinformatics. There is no way to predict exactly how these will affect developing-country agriculture, but they will all add to the foundation of knowledge on which scientific and technological discoveries are made.

Bioremediation: Research in both natural and physical sciences has shown that plants and microbes can be used to remove contaminants from the environment. Bioremediation techniques are being developed to clean up oil spills, hazardous wastes, and other pollutants. Enhancing the biocatalytic characteristics of some plants would be valuable in particular developing regions where harsh environments, depleted resources, or unusual habitats preclude production with current technologies.

Nanotechnology: The development of microscopic tools for imaging and manipulating single molecules has led to the exciting new field of nanotechnology. Ultra-small structures and machines are being made of as few as one molecule. Bio-nanotechnology may give molecular biologists even greater opportunities to investigate the physiological functions of plants and animals, which can increase the speed and power of disease diagnosis.

Genomics: Genomics is the study of the genome and the biological roles genes play, individually and collectively, in determining structure, directing growth and development, and controlling biological functions. Public and private projects have generated genome maps and complete deoxyribonucleic acid (DNA) sequences of several organisms. Two biotechnology companies donated research results to the international effort to produce a complete genetic map of rice.

Genetic sequence information can be used to develop diagnostic tests, find genetic markers, identify genetic susceptibilities, and develop therapeutics. The role genes play in biological functions involves protein production. Genes exert their effects through proteins, but less is known about the link between proteins and biological function. Proteomics is the study of the structure, function, location, and interaction of proteins within and between cells.

Bioinformatics: This technology uses statistical software, graphics simulation, and database management to consistently organise, access, process, and integrate data from different sources. Specific activities may include screening chemical compounds, identifying potential pharmaceutical drugs, and determining plant and animal genes to improve sustainable agricultural production. Bioinformatics has already been used to form international databases that are available to scientists around the world via the Internet. In this way, the quality of the data on plants, animals, and microbes can be assessed, and the information made accessible to researchers in both developed and developing countries.

Technology Development and Transfer

In any particular country, a variety of economic, social, environmental, and institutional factors can create high barriers to technology development and transfer. These factors will be discussed under broad headings:

— Systems for providing scientific research, public health, and education,

— Economic infrastructure, such as transportation and communications networks,

— Financial, legal, and political institutions, including intellectual property rights,

— Natural resources and environmental regulations, and

— International treaties and trade policies.

Support for technology development and transfer includes contributions to reducing barriers and increasing incentives.

Research, Health, and Education Capacity

Research systems-both public and private — play a critical role in developing new productivity-enhancing agricultural technology, as well as in facilitating technology transfer and adaptation to developing countries. Scientists with knowledge of local crops and environments are crucial for ensuring the selection and development of appropriate technologies. Issues of finance and governance are relevant to the performance of agricultural research, just as they are in other areas of public investment or public policy.

Developing political support for public sector agricultural research, finding the means of financing such research, and setting priorities that are reflected in the allocation of research budgets are important policies that support technology development and transfer. Most low-income countries do not have large financial resources to invest in the training of scientists, maintenance of research facilities, or many other components of a strong agricultural R&D program.

Asian countries have been able to invest more than most countries in Africa, as is reflected by the relative levels of per capita food production in the two regions. The average level of agricultural R&D expenditure in Asia, however, is still below the world average. Since resources may not be available for domestic investment in research, there is a need to transfer technologies to increase agricultural productivity and income.

But technology transfer entails more than just shipping machines, seeds, or blueprints. Experts with knowledge of their country's characteristics are needed to adapt technologies and to develop effective incentives to ensure adoption and efficient use. Research infrastructure in the poorest regions can be improved through direct investment in facilities and education in the developing country, and through the support of such organisations as the World Bank, the Rockefeller Foundation, and the Consultative Group on International Agricultural Research (CGIAR).

International collaboration in public agricultural research has been very successful in transferring basic and applied knowledge throughout the world. Complementary public investments may also influence the success of agricultural research and technology transfer.

A strong public health system is important to the success of new agricultural technology and to the development of agriculture generally. If the agricultural labor force is in poor health, it will be much more difficult to raise agricultural productivity with or without new technology. Malaria, tuberculosis, and other chronic diseases as well as the prevalence of micronutrient deficiencies compromise food and nutrition security.

Education is important at all levels to support the development and transfer of science and technology. In addition to the scientists directly involved in research, there is a need for trained individuals to develop and implement regulations that affect technology use. Qualified people are also needed to represent their country's interests in international negotiations. Decision-makers need the expertise to understand the positive and negative implications of their actions within the complex human and ecological environment of their country.

Labor quality may vary with differences in experience and education, making investment in basic education another potential complement to investment in agricultural research. Particularly for knowledge-intensive technologies such as precision agriculture, farmer education may be crucial to adoption. It is important that educational opportunities be nondiscriminatory, for example on gender or ethnic grounds. Women often make the agricultural production decisions, and their knowledgeable input into technology choices is essential.

Economic Infrastructure

Economic infrastructure includes services from public utilities such as power, telecommunications, water supply, and sanitation and sewerage; and from public works and transport systems such as dams and canals for irrigation and drainage, roads, railways, airports, ports, and waterways. Infrastructure can be highly complementary to science and technology in increasing agricultural productivity, and lack of infrastructure can seriously constrain agricultural development.

For example, those countries that had poor transportation, irrigation, and financial systems did not benefit from the Green Revolution technology as much as those countries that did. Green Revolution technology was most closely associated with new crop varieties that were combined with increased use of fertilizer in areas with irrigation or more rainfall. The countries in which highyielding varieties were most successful also had functioning roads, irrigation systems, options for credit, and markets and distribution channels.

A functioning agricultural market is dependent on a strong communication infrastructure. Rural areas rely on public information to be integrated into the national economy. Basic communication that needs to be made available includes information about health and sanitation hazards, weather, public transportation schedules, labor and market opportunities, and rights to public resources. Information from scientists and agricultural experts about production practices needs to be disseminated as widely as possible.

Moreover, as farmers vary production to include horticulture, vegetables and fruits, and other products that must be delivered quickly, there is a premium on real-time information about markets. Information and communications technology has been transformed throughout the world, by computer and Internet use and by wireless telephone technology. These innovations can help farmers in all countries by providing upto-date market and labor information. These innovations can also serve to provide educational opportunities and timely transmission of critical information.

For those farmers in developing countries with Internet access, agricultural extension information and research results can be disseminated widely. Farmers could also alert researchers to emerging pest pressures and environmental conditions. To make this technology more effective in smallholder agriculture, however, literacy and physical infrastructure, such as electrical power, must be as widely accessible as possible.

Institutional Infrastructure

The introduction of new technologies requires new policies to ensure health, safety, or environmental quality. Stable institutions are important for long-term planning and investment in technology development. Open and transparent investment regulations that are compatible with global trading rules will encourage both domestic and foreign investment. Perceived fairness will also encourage participation at all levels, which will encourage foreign investment as well as farm-level cooperation with technology development plans.

Well-functioning markets that operate internationally and locally depend on the strength of a country's financial, legal, and political institutions. Private investment in technology development and transfer from domestic and foreign sources will not be forthcoming without a strong demand by farmers and a well-functioning infrastructure. Financial institutions provide capital for research, physical infrastructure, and farm credit. Devising agricultural credit systems that fit the needs of smallholders has been difficult in many countries, especially those with large income gaps, complex land ownership, weak banking systems, and under-employment.

This is an important barrier because many new agricultural technologies require the use of purchased equipment or inputs, making credit essential to their widespread use. Legal institutions often reflect a country's social history. Rights and private asset ownership will determine who has access to institutional assets such as education, finances, or the right to participate in the political process. For agricultural producers in developing countries, control of land is often a critical determinant of technology adoption.

Investment in agricultural technology is often related to a farmer's security of land tenure. In addition, natural resource conservation efforts with long-term benefits may be hampered without land tenure security. As mentioned above, contracts that could facilitate better access to inputs or improved markets need to be legally enforceable. Political institutions need to be stable to support opportunities for agricultural development.

Yet the political environment of a country is more than just the formal deliberations of a national leader or a legislative body. It includes the planning, administrative, regulatory, and enforcement functions that are often dismissed as "merely bureaucratic." These functions, if performed efficiently, are critical to the research, development, dissemination, and adoption of science and technology.

Development planning must have the support of government leaders and the general population so that agricultural production and environmental quality goals reflect the true needs of the country. The benefits of any development plan will increase when science and technology policies are integrated in the plan, and when there is a framework to implement the decisions. Technical support for planning from other countries may be needed, but the goals and objectives must be national.

In support of planning and implementation, there is a need for a regulatory process that is designed to protect the health and well-being of the people and the environment. The assessments of new technologies need to be done quickly and thoroughly. Domestic expertise that is fostered by strong research and education capacity will be needed to make decisions that are based on sound science and that meet the needs of the country. Regulatory frameworks and testing protocols from other countries can be used as models. Domestic scientists and technical experts, however, are needed to monitor and adapt implementation to regional circumstances.

In many countries, a tension exists between agricultural and environmental interests, even within the government. With increased understanding of the complex interactions between agricultural production and environmental assets, opportunities arise to develop a science and technology plan that supports a country's agricultural development and environmental quality goals. In some countries, there is no regulatory system within which science-based decisions can be made. Without such a system, potentially beneficial innovations and technologies are not being considered for investment or adoption.

Several specific components of the institutional infrastructure are particularly important for understanding barriers to science and technology transfer: A nation's system of intellectual property rights, access to germplasm for research, and the body of domestic agricultural policies will all influence relative prices and incentives for public and private research.

Intellectual Property Rights (IPR)

This is one set of legal rights of particular importance to research and technology transfer. The critical issue in respect of IPR is perhaps not whether it promotes trade or foreign investment, but how it helps or hinders developing countries to gain access to technologies that are required for their development. Currently, the lack of comprehensively specified and enforceable intellectual property rights constitutes a major barrier to the sharing of knowledge and technology among countries, and a disincentive to local and foreign research investment in new technologies.

In general, nations that generate technology prefer strong intellectual property protection, while those that depend on imported technologies prefer few restrictions on the use and imitation of that technology. Over the past 20 to 30 years, intellectual property systems have become increasingly important factors affecting research in industrialised countries. Intellectual property mechanisms, such as patenting, confer exclusive rights to inventions for a limited time, to offer an incentive for research and to protect private sector investment in new technologies and products by restricting the use, sale, and manufacture of these innovations.

Research investment can be costly, and the probability of success low. Many firms would not risk funding research if they were unable to obtain a return on the investment. Therefore, new technologies become available earlier than they would without intellectual property protection. While intellectual property protection has been a feature of chemical and mechanical innovation in industrialised agriculture for some time, changes in intellectual property protection that have had the greatest impact on agricultural R&D have involved biological innovation.

The two major forms of protection are plant varietal protection, or plant breeders' rights, and the application of utility patents both to plants and to biological research tools. Although intellectual property protection may speed some inventions to the market, exclusive rights to fundamental innovations may impede further technological progress. Often, technology development is cumulative, and scientific advances depend on past innovations.

Restrictions on the use of innovations for research could limit future research or needed adaptations of past inventions. This barrier to research is an issue in developed as well as developing countries. Partnerships between public institutions and private companies often include provisions to grant access to public research results. Many are concerned that private commercial research will not be done on crops that are important for local staple crops, and that public research entities will not have access to the basic discoveries necessary to develop technologies to fill smallholder needs in developing countries.

Germplasm access

Crop improvement through traditional plant breeding methods and modern biotechnology depends critically on crop genetic resources. Pests, pathogens, and climates change continually, so breeders need new genetic resources from which to choose desired traits. For example, germplasm is used in a research program to search for resistance to or tolerances of biotic stresses. Even though many sources of germplasm are located in developing countries, the usefulness of these resources to these countries depends on research funding and infrastructure to utilize the materials.

The International Plant Genetic Resources Institute also shares genetic resources with the scientific community, and has helped to establish over 1,300 national and regional genebank collections. Enhanced availability of genetic resources and increased indigenous research capacity will make it easier for crops to be developed to meet the unique needs of each developing country. Historically, plant genetic material was freely collected and shared. Developing countries-with a wealth of biological diversity in situ (in the wild and on fields)-often provided raw genetic material to public genebanks worldwide.

However, international policy has moved toward a system in which countries retain rights over their own genetic resources, and the services of farmers in the selection, development, and conservation of their traditional varieties-the foundation on which plant breeding is based-are better recognised. The goals of granting national ownership to genetic resources were to provide incentives for the conservation of diverse germplasm and to address perceived economic inequities between suppliers and demanders of germplasm.

This new policy approach is represented by the International Treaty on Plant Genetic Resources for Food and Agriculture, which the U.N. Food and Agricultural Organisation approved in November 2001 and which now awaits ratification. This international treaty will govern international exchange of germplasm among countries participating in a multilateral system. Issues of particular interest to developed and developing countries that remain to be resolved, however, include the implementation of benefit sharing, financing conservation, and the list of crops in the system.

Domestic agricultural policies

In many developing countries, the agricultural sector makes an important contribution to the gross national product (GNP). However, domestic policies often penalize agriculture and distort markets, and may be counterproductive in the long run. Farmer and investor choices of technologies and practices are based on prices and costs. If these economic signals are distorted by fiscal or monetary policies, the technology adoption incentives will not be optimal. Market price supports, direct payments to farmers, input subsidies, agricultural taxes, or monetary and trade policies can mask the "true" prices and costs of production and product, thus distorting incentives for the adoption of technology.

Natural Resource Base and Environment

A country's natural resource base and environment are crucial factors in determining realistic sustainable agriculture development goals. Although climate and natural disasters are not under a planner's control, the vulnerability to these factors can be mitigated. There are striking regional differences in cropland quality. Among the countries of Sub-Saharan Africa, an average of 6 percent of cropland has soils and climate that are of high quality for agricultural production.

The proportion of high-quality cropland is higher in other regions, ranging from an average of 20 percent among Asian countries, to 28 percent among the countries of Latin America and the Caribbean, and 29 percent among high-income countries. In countries with poor soils and climate, basic inputs such as fertilizer and water are more important than they are in countries that are better endowed.

Given economic and environmental constraints on cropland expansion, the bulk of increased crop production in the future must come from increased yields on existing cropland. In some areas, yield increases may be constrained by soil ero-sion and other forms of land degradation. Recent U.S. Department of Agriculture studies show that yield losses (or lack of gains) due to soil erosion vary widely by crop and region, and the losses critically depend on the agricultural practices that are used.

To lower these losses, improvements in resource-conserving technologies in some developing countries, and in incentives to farmers to use appropriate practices, may be needed. In developing countries, increased yields have come at the cost of negative environmental impacts. These impacts may include water pollution, salinisation and land abandonment, lowering of groundwater levels, and loss of biodiversity with more uniform crops.

However, large areas of environmentally fragile land, which might have been pressed into production had yields not increased in the more favored areas, were saved. Nonetheless, in recent years many developing countries, including those that have benefited the most from the Green Revolution, have been showing signs of a slowdown in agricultural productivity gains.

At least part of this slowdown might be attributed to environmental problems related to intensive agriculture. Therefore, greater research emphasis on environmental concerns is increasingly important.

International Agreements and Policies

The effectiveness of a country's science and technology policy will depend, to some extent, on international agreements and policies. These international policies are largely out of the control of a single nation, but domestic policy development must take them into account. The complexity of many of these agreements, however, makes it difficult to accurately assess the extent and timing of impacts.

World Trade Organisation

World Trade Organisation (WTO) is a multilateral institution charged with administering transparent rules for global trade among member countries. The WTO fosters trade liberalisation that, in turn, brings about important transfers of technology, capital investment, and knowledge across borders. The WTO was established in 1995 as a result of the Uruguay Round, where countries agreed to initiate a more fair and market-oriented agricultural trading system. At the 4th Ministerial in Doha in 2001, WTO members engaged in new multilateral trade negotiations. For agriculture, the Doha Declaration calls for substantial improvements in market access, and the reduction of all forms of export subsidies and trade-distorting domestic support.

The Agreement on Trade-Related Intellectual Property Rights (TRIPS) was part of the WTO negotiations, and covers patents, copyrights, trademarks, industrial designs, plant varieties, and trade secrets. Several developing countries expressed the hope that scientific and technological cooperation between developed and developing countries in accordance with the provisions of the TRIPS Agreement would support public interest issues such as health, nutrition, environmental protection, and natural resource conservation in developing countries.

Industrialised countries and international organisations were asked to help developing countries implement the TRIPS Agreement by 2006, but national expertise is needed to weigh the benefits and costs of each option. TRIPS will affect the science and technology that will be available to increase productivity.

The Sanitary and Phytosanitary (SPS) Agreement established a multilateral mechanism to protect human, animal, and plant health in WTO member countries. SPS measures are required to be based on scientific principles, and the nature and magnitude of the perceived risk must be clearly established. Technologies and practices used in the production of agricultural commodities for trade may be restricted under SPS rules. The science- and research-based requirements of the SPS Agreement can be substantial for a developing country.

Biosafety Protocol

The Biosafety Protocol to the United Nations Convention

on Biological Diversity provides a regulatory framework for transboundary movements of living bio-engineered organisms. The Protocol requires that regulatory decisions to deny entry of a product in order to avoid or minimise potential adverse effects must be based on risk assessments and sound science. The importation and use of some biotechnology applications may be affected. The Protocol also establishes biosafety clearinghouse to help countries assess potential risks from genetically engineered organisms.

This provision may be useful in addressing the concerns of many countries that believe their current regulatory systems are inadequate deal with the potential implications of the technology on the nation's environment. Even with a clearinghouse in place, many developing countries will need local expertise to make knowledgeable decisions consistent with national sustainable agriculture goals.

Scientific breakthroughs and technological innovations in the 20th century fueled substantial gains in agricultural productivity in many developed and developing countries. The development of new technologies and practices resulted from both public and private investments in research. Countries that enjoyed high agricultural productivity growth were able to increase incomes, participate in global markets, reduce hunger and poverty, and improve the quality of life of their citizens.

For the countries that were not able to benefit from the advances in science and technology, agricultural productivity did not grow quickly, which resulted in unmet needs for income growth and food security. Many technologies and practices developed in the 20th century could be adapted to meet the unique needs of each developing country. Scientific understanding about the interactions between agricultural production and ecosystem health can also contribute to the development of a sustainable agricultural system. The choice of an appropriate set of technologies and practices should incorporate indigenous knowledge of the local economic, social, and natural resource environment.

Agricultural production technologies and practices have been developed to improve soil, water, nutrient, and pest management. Crop improvements contributed to the successes of the Green Revolution. Tools of modern biotechnology have been used to achieve higher levels of stability and sustainability in crop production. These innovations have increased yields and reduced environmental impacts.

Advances in animal breeding and health have increased both the quantity and quality of animal protein available to consumers. Improvements in marketing, processing, and transportation technologies have expanded the choices of food that are readily available to consumers in developed economies. These innovations can be adapted to preserve and deliver vitaminrich foods to help combat nutrient deficiencies in developing countries.

In addition, technologies to reduce food safety hazards can be used to increase the health of both rural and urban populations. Scientific and technological advances in the 21st century will result from research investments in both traditional agricultural fields and other emerging disciplines. Agricultural production research will be targeted to develop crops and animals that can tolerate a wider range of environmental conditions and offer consumers desired characteristics.

Molecular methods will be used to diagnose diseases, locate pollutants in the environment, and detect harmful micro-organisms in food. Modern biotechnology holds promise for the production of pharmaceutical compounds such as vaccines within locally grown plants. Innovations in biological and information sciences have resulted in several emerging fields that hold promise for the development of future agricultural technologies. The new fields of bioremediation, nanotechnology, genomics, proteomics, and bioinformatics will increase knowledge that can be shared and used to improve sustainable agricultural production and protect ecosystem functions in developed and developing countries alike.

Scientific and technological advances hold great promise, but the full benefits of scientific breakthroughs will not be realised without the dissemination and adoption of new technologies. In each country, the successful local development of technologies or the transfer and adaptation of innovations from other countries will depend on incentives and barriers faced by investors and producers.

Countries with strong research, health, and education capacity will offer a supportive environment for technology investment. Financial resources are needed to train scientists, enhance and maintain research facilities, develop agricultural markets, and provide adequate health and education systems to the population. External funds could be used to fund these efforts, but the priorities for development must come from within developing countries to ensure that their unique economic, social, and environmental needs are met.

Inadequate public utilities, transportation systems, and other infrastructure will impede the development of agricultural markets by limiting the availability of affordable inputs and inhibiting the timely delivery of highquality agricultural products. Financial, legal, and political institutions have profound effects on technology development and transfer and on the evolution of agricultural markets. Incentives for domestic and foreign investment are tied to the stability and perceived fairness of the institutional infrastructure within a country. Domestic agricultural policies within developing countries may affect prices and costs, thus distorting incentives for research investment and technology adoption.

The sharing of knowledge between countries currently is hindered in some cases by intellectual property rights systems that differ between countries. Innovative public/private partnerships are being designed to help developing countries gain access to new technologies. It may be difficult to achieve development goals for a sustainable agricultural system in countries that have a poor natural resource base or an environment that is vulnerable to degradation.

These conditions limit the choices of technologies and practices that are appropriate to use. In addition to domestic circumstances, international treaties and trade policies have impacts on the success of science and technology policy in developing countries. The liberalisation of global trade will affect prices and incentives to invest.

The TRIPS Agreement has the potential to enhance the science and technologies that will be available to increase agricultural productivity, and the Biosafety Protocol contains provisions for a clearinghouse to help developing countries make sciencebased decisions about trade in bio-engineered products. International deliberations can have an effect on decisions regarding investment in technology, even if the country does not actively participate in global markets.

Increasingly, research investments and technology transfer will depend on cooperative endeavors between developed and developing countries and between public and private institutions. Developing countries have many crucial decisions to make in meeting their goals for sustainable agricultural systems. These decisions need to be made and implemented based on the knowledge of each country's unique environmental, social, and economic characteristics.

Local expertise is needed to take advantage of indigenous knowledge, and to establish environmental and food safety safeguards to ensure that both the positive and negative potential impacts of a new technology are adequately assessed. There are many ways that developed countries, international institutions, and businesses can increase the possibilities for developing countries to benefit from scientific and technological advances.

They can continue to train scientists and offer the expertise needed to help develop workable plans to achieve productive and sustainable agricultural systems. Investment incentives can be increased directly, and by helping developing countries establish and maintain the legal, financial, transportation, and communications infrastructure necessary to encourage investment.

Public and private investment in research to increase agricultural productivity of the poorest nations can have many benefits. With supportive policy, regulatory, and institutional frameworks in place, science and technology can drive agricultural productivity increases, alleviate hunger, and foster economic growth in developing countries. Incentives for private investment will increase as regions gain the economic resources to participate more actively in the global marketplace.

Higher incomes and better nutrition will improve food security and allow more developing countries to share in the growth that many countries have enjoyed for the past half century. Thus, science and technology can play a critical role in helping to prevent famine, improve nutrition, and move countries closer toward a goal of ending world hunger.

4

Transducer Fundamentals

Oxygen Electrode and Amperometry

Amperometry

Polarogram: When an electrode of noble metal such as platinum or gold is made 0.6 to 0.8 V negative with respect to a suitable reference electrode such as Ag/AgCl or an calomel electrode in a neutral KCl solution (Fig. 1), the oxygen dissolved in the liquid is reduce at the surface of the noble metal.

This phenomenon can be observed from a current-voltage diagram - called a polarogram - of the electrode. As shown in Fig. 2a, the negative voltage applied to the noble metal electrode (called the cathode) is increased, the current increases initially but soon it becomes saturated. In this plateau region of the polarogram, the reaction of oxygen at the cathode is so fast that the rate of reaction is limited by the diffusion of oxygen to the cathode surface.

When the negative bias voltage is further increased, the current output of the electrode increases rapidly due to other reactions, mainly, the reduction of water to hydrogen). If a fixed voltage in the plateau region (for example, - 0.6V) is applied to the cathode, the current output of the electrode can be linearly calibrated to the dissolved oxygen (Fig. 2b). It has to be noted that the current is proportional not to the actual concentration but to the activity or equivalent partial pressure of dissolved oxygen, which is often referred to as oxygen tension. A fixed voltage between -0.6 and -0.8 V is usually selected as the polarization voltage when using Ag/AgCl as the reference electrode.

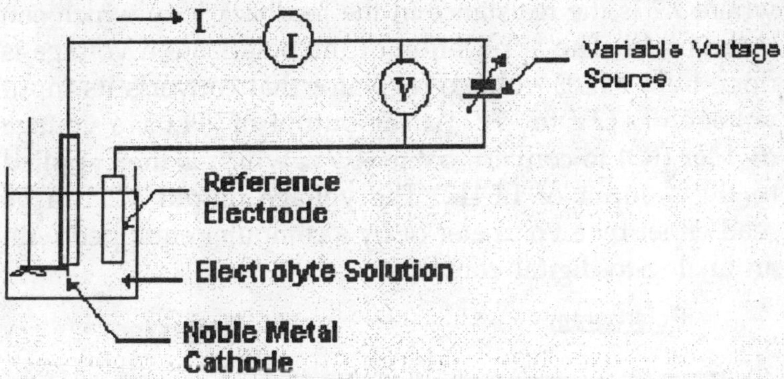

Figure 1. Setup for polarography

DO Sensor: When the cathode, anode, and the electrolyte are separated from the measurement medium by a polymer membrane, which is permeable to the dissolved gas but not to most of the ions and other species, and when most of the mass transfer resistance is confined in the membrane, the electrode system can measure oxygen tension in various liquids. This is the basic operating principle of the membrane covered polarographic DO probe (Fig 3).

Signal Conditioning: To read the output from the sensor, the current from the sensor is converted to voltage by the circuit shown in Fig. 4 (the first 1/2 of an operational amplifier LF412). This circuit has a gain of 10,000,000:

$$V_1 = - I * 10{,}000{,}000$$

Therefore, 0.1 µA sensor current will produce an output of - 1V at pin 1 of LF412 (note that R1 can be changed to obtain other amplifier gains). The next stage is an inverting amplifier with gain. The output from this stage is:

$$V_2 = -V_1 \times (R2/1000) = 10{,}000 \times R2$$

where R2 is the resistance in the feedback loop which can be adjusted. The application of the polarization voltage is done by a 79L05 voltage regulator that converts its input voltage of -12V to -5V. At the output of 79L05, a voltage divider (R3) to convert -5V to -0.7V, which is then applied to the + input of LF412. The voltage output V2 can be read either by a voltmeter or by a computer equipped with an analog-to-digital converter.

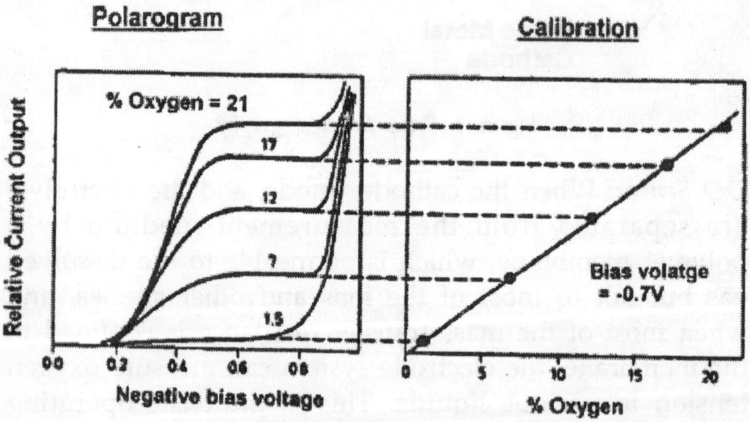

Figure 2. (a) Current-to-voltage diagram at different oxygen tensions; (b) calibration obtained at a fixed polarization voltage of -0.6V.

Electrode For polarographic electrodes, the reaction proceeds as follows:

Reactions

Cathodic reaction: $O_2 + 2H_2O + 2e^- \rightarrow H_2O_2 + 2OH^-$

$$H_2O_2 + 2e^- \rightarrow 2OH^-$$

Anodic reaction: $Ag + Cl^- \rightarrow AgCl + e^-$

Overall reaction: $4Ag + O_2 + 2H_2O + 4\ Cl^- \rightarrow 4\ AgCl + 4\ OH^-$

the reaction tends to produce alkalinity in the medium together with a small amount of hydrogen peroxide.

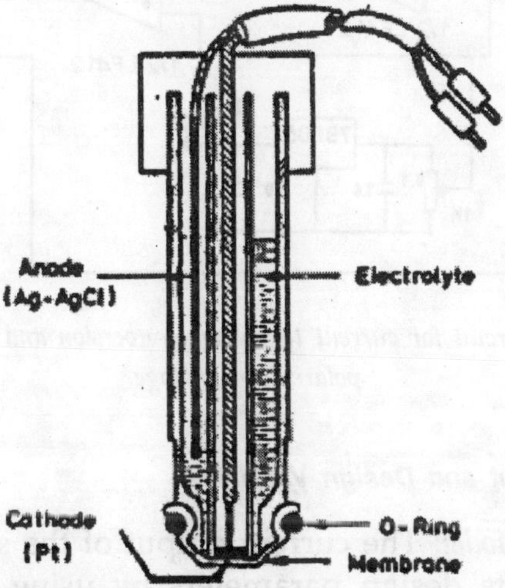

Figure 3. Membrane covered polarographic oxygen sensor.

Electrons Involved: Two principal pathways was proposed for reduction of oxygen at the noble metal surface. One is a 4 electron pathway where the oxygen in the bulk diffuses to the surface of the cathode and is converted to H_2O via H_2O_2 (path a in Fig. 5). The other is a 2 electron pathway where the intermediate H_2O_2 diffuses directly out of the cathode surface into the bulk liquid (path b in Fig. 5).

The oxygen reduction path may change depending on surface condition of the noble metal. This is probably the cause for time-dependent current drift of polarographic probes. Since the hydroxyl ions are constantly being substituted for chloride ions as the reaction starts, KCl or NaCl has to be used as the electrolyte. When the electrolyte is depleted of Cl^-, it has to be replenished.

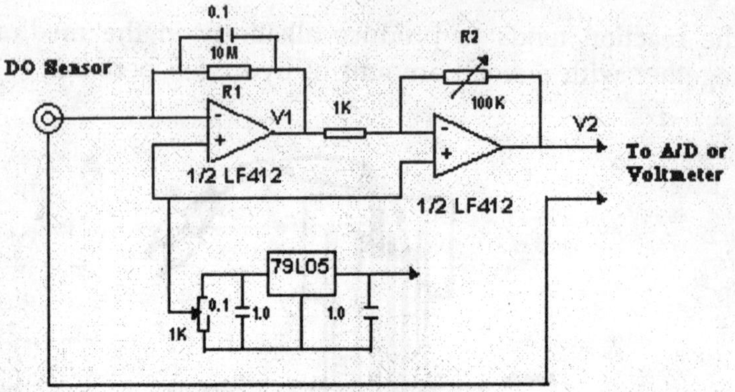

Figure 4. Circuit for current to voltage conversion and application of polarization voltage.

Sensor Output and Design Variables

One Layer Model: The current output of the sensor can be related to its design parameters by using a simplified electrode model. The assumptions used for developing the model:

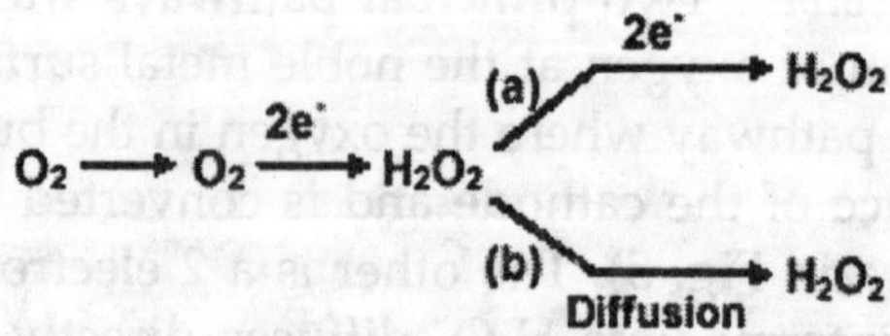

Figure 5. Alternative pathways of oxygen reduction at cathode surface.

Assumptions Used

— The cathode is well polished and the membrane is tightly fit over the cathode surface such that the thickness of the electrolyte layer between the membrane and the cathode is negligible.

— The liquid around the sensor is well agitated that the partial pressure of oxygen at the membrane surface is the same as that of the bulk liquid.

— Oxygen diffusion occurs only in one direction, perpendicular to the cathode surface.

1-D Diffusion Equation: This so-called one layer model can be extended to include the effects of other layers (such as the liquid layer) as will be shown later. According to Fick's 2nd law, the unsteady-state diffusion in the membrane is described by (using the coordinate system shown in Fig.6):

$$\frac{\partial p}{\partial t} = D_m \frac{\partial^2 p}{\partial x^2} \tag{1}$$

where D_m is the oxygen diffusivity in the membrane and x is the distance from the cathode surface.

Boundary Conditions: The initial and boundary conditions are:

$$p = 0 \quad \text{at } t = 0 \tag{2}$$

$$p = 0 \quad \text{at } x = 0 \tag{3}$$

$$p = \mathrm{p_o} \quad \text{at } x = d_m \tag{4}$$

where d_m is the membrane thickness and po is the partial pressure of oxygen in the bulk liquid. The second boundary condition (Eq. (3)) assumes a very fast reaction at the cathode surface, which is generally achieved when the cathode is properly polarized (with the right voltage).

Unsteady Oxygen Profile: The solution of Eq. (1) with the boundary conditions is:

$$\frac{p}{p_o} = \frac{x}{d_m} + \sum_{n=1}^{\infty} \frac{2}{n\pi} (-1)^n \sin \frac{n\pi x}{d_m}$$

$$\exp(-n^2 \pi^2 D_m t / d_m^2) \quad (5)$$

The current output I of the electrode is proportional to the oxygen flux at the cathode surface:

$$\mathrm{I} = NFAD_m(\partial C/\partial x)_{x=0} \quad (6)$$

$$= NFAP_m(\partial p/\partial x)_{x=0}$$

where N, F, A, and P_m are the number of electrons per mole of oxygen reduced, Faraday constant (= 96,500 coul/mol), surface area of the cathode, and oxygen permeability of the membrane, respectively. The permeability P_m is related to diffusivity by:

$$P_m = D_m S_m \quad (7)$$

where S_m is the oxygen solubility of the membrane.

Unsteady Current Output From Eqs. (5) and (6), the current output It of the electrode is:

$$I_t = NFA \frac{P_m}{d_m} p_o \left[1 + 2 \sum_{n=1}^{\infty} (-1)^n \exp(-n^2 \pi^2 D_m t / d_m^2) \right] \quad (8)$$

Steady Current Output The pressure profile within the membrane and the current output under steady-state conditions can be obtained from Eqs. (5) and (8), respectively:

$$\frac{p}{p_o} = \frac{d}{d_m} \tag{9}$$

$$I_s = NFA\frac{P_m}{d_m}p_o \tag{10}$$

At steady state, the pressure profile is linear and the current output is proportional to the oxygen partial pressure in the bulk liquid. Eq. (10) forms the basis for DO measurement by the sensor.

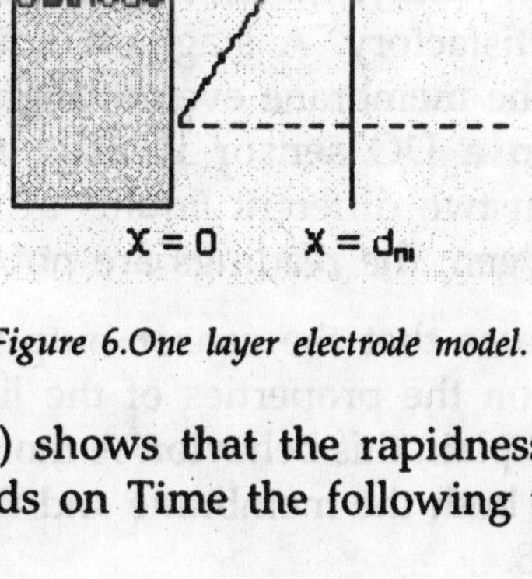

Figure 6.One layer electrode model.

Response: Eq. (8) shows that the rapidness of the sensor response depends on Time the following term:

$$\tau = \frac{d_m{}^2}{D_m} \tag{11}$$

When τ is large (a thin membrane and/or high D_m), the sensor responds fast. Note that these conditions tend to weaken the assumption of membranecontrolled diffusion. Therefore, a compromise has to be made for an optimum sensor performance. Note that adjusting d_m (rather than D_m) is more effective in adjusting τ (because τ depends on the square of d_m). Eqs. (10) and (11) indicate that the design variables for a DO sensor is P_m, d_m, D_m, and A (Fig.7).

Current Output: $I_s = NFA\dfrac{P_m}{d_m}P_0$

Response Time: $\tau = \dfrac{d_m{}^2}{D_m}$

Design Variable: P_m, d_m, D_m, A

Figure 7. Design variables of DO sensor.

Liquid Velocity and Properties on Sensor Performance

Real Situation: In reality, the assumption 2 made earlier is not entirely satisfactory. A stagnant liquid film always exists outside the membrane even at high liquid velocity. Actually, when a DO sensor is used to measure DO concentration in two different liquids at the same partial pressure of oxygen, the readings are not the same.

This indicates that the sensor output depends, to a certain extent, on the properties of the liquid. One layer model cannot explain this behavior. A more realistic model has to consider both the membrane and the liquid film as shown in Fig. 8.

Two Layer Model: The effect of liquid layer on sensor current output can be estimated by expanding the "one layer" model. At steady state, the oxygen flux J through each layer in Fig. 8 should be the same:

$$
\begin{aligned}
J &= Kp_o \\
&= kL(p_o - p_m) \qquad (12) \\
&= k_m p_m
\end{aligned}
$$

where K is the overall mass transfer coefficient and kL and km are mass transfer coefficient for liquid film and membrane, respectively.

Ohm's Law Analogy: The inverse of the mass transfer coefficient can be termed as the mass transfer resistance. From Eq. (12), it can be shown that:

$$\frac{1}{K} = \frac{1}{k_L} + \frac{1}{k_m} \qquad (13)$$

Equation (13) says that the overall mass transfer resistance, $1/K$, is the sum of the liquid phase mass transfer resistance, $1/k$, and the membrane phase mass transfer resistance, $1/k_m$. The derivation is based on Ohm's law analogy - J is considered the current and p as the voltage. The individual resistances can be replaced by:

$$\frac{1}{K} = \frac{d_L}{P_L} + \frac{d_m}{P_m} \qquad (14)$$

where d_L and P_L are liquid film thickness and the oxygen permeability of the liquid film, respectively. Note that a stagnant liquid film was assumed here, although it is more accurate to use the convective mass transfer coefficient.

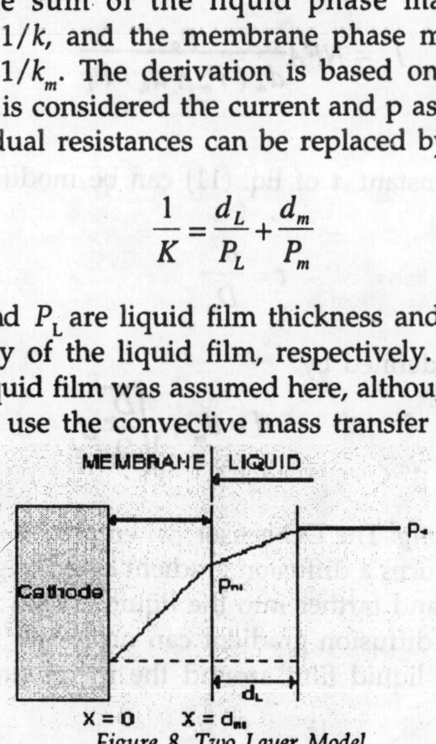

Figure 8. Two Layer Model

Two Layer Model: When the individual mass transfer resistances are considered the steady state sensor output becomes:

$$I_s = NFA\frac{P_m}{\bar{d}}p_o \tag{15}$$

where $\bar{d}$ is defined by:

$$\bar{d} = d_m + P_m\frac{d_L}{P_L} \tag{16}$$

Alternatively, I_s can be written as:

$$I_s = NFA\frac{p_o}{d_m / P_m + d_L / P_L} \tag{17}$$

The time constant τ of Eq. (11) can be modified to:

$$\tau = \frac{d^2}{D_m} \tag{18}$$

where d is defined by:

$$d = d_m + \sqrt{\frac{D_m}{D_L}}d_L \tag{19}$$

Flow Sensitivity: The DO sensor, when placed in a stagnant liquid, produces a diffusion gradient extending outside the membrane and farther into the liquid. When the liquid is stirred, the diffusion gradient can no longer be extended beyond the liquid film around the membrane. Since the diffusion gradient becomes steeper with decreasing liquid film thickness, the current output of the sensor increases with increase in liquid velocity (Fig. 9a).

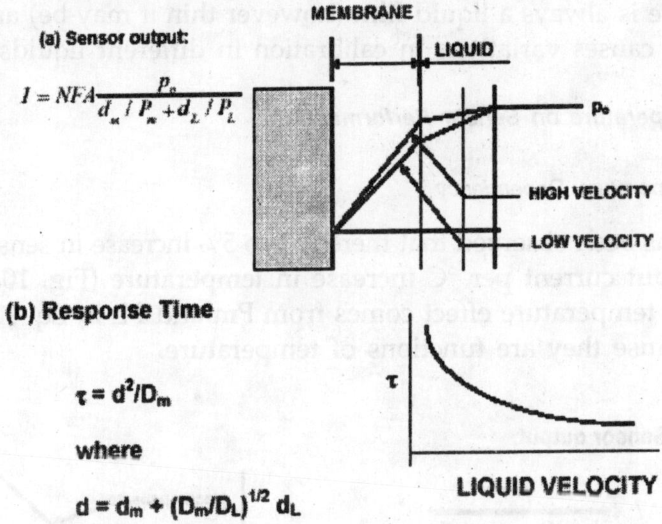

Figure 9. Flow sensitivity of DO sensor

Note also that the response time of the sensor increases as the liquid velocity decreases (Fig. 9b). This so-called "flow sensitivity" is greater for a sensor with a larger cathode because the size of the stagnant diffusion field is proportionally greater with a larger cathode.

Condition for Membrane Control of Diffusion: From Eq. (17), the condition for a membrane-controlled diffusion becomes:

$$d_m/P_m >> d_L/P_L \tag{20}$$

To achieve this condition, a relatively thick membrane with a low oxygen permeability have to be used. Note that this contradicts the requirement for a fast sensor response. When this condition is achieved, the oxygen sensor output depends only on membrane properties as given by Eq. (10) and the sensor calibrated in one liquid can be used in other liquids without recalibration. In reality, however, there is always a liquid film (however thin it may be) and this causes variations in calibration in different liquids.

Temperature on Sensor Performance

Temperature Dependency

It has been observed that there is 1 to 5% increase in sensor output current per °C increase in temperature (Fig. 10a). The temperature effect comes from Pm and PL in Eq. (17) because they are functions of temperature.

(a) Sensor output:

$$I = NFA\frac{P_o}{d_m / P_m + d_L / P_L}$$

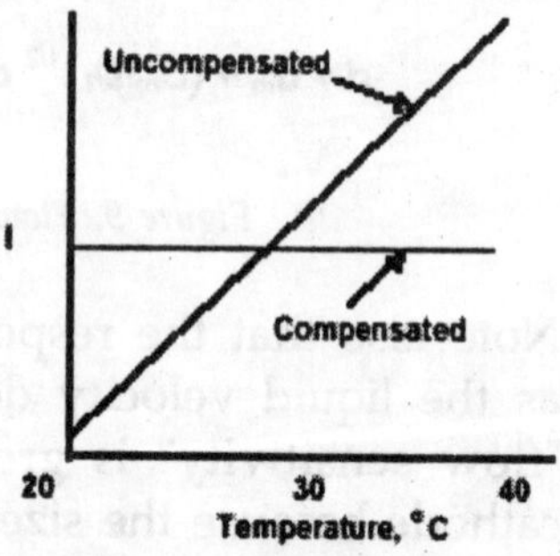

(b) Response Time

$$\tau = d^2/D_m$$

where

$$d = d_m + (D_m/D_L)^{1/2}\, d_L$$

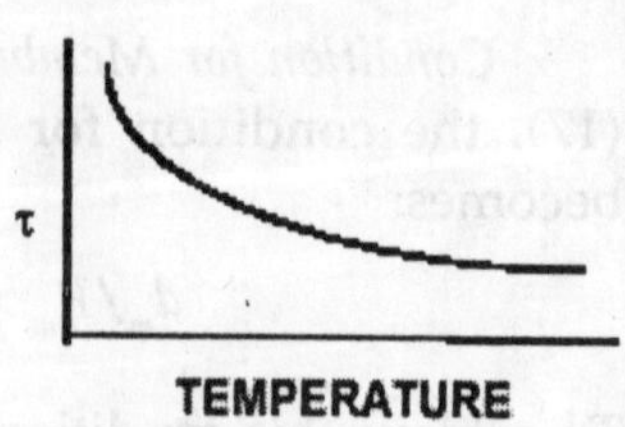

Figure 10. Temperature dependency of DO sensor.

When the sensor is operated under membranediffusion control, the temperature dependency of I_s should come entirely from P_m whereas when the liquid film resistance is not negligible, both P_m and P_L contribute to the temperature dependency. Generally, P_m is expressed as:

$$P_m = P_m^* \exp(-E/RT) \quad (21)$$

where E is the activation energy for the permeation and amounts to be 8.8 kcal/g-mol for polyethylene and 7.8-9.6 kcal/g-mol for polypropylene membranes. The response time of the sensor also depends on temperature but the temperature dependency originates from diffusivity (Eqs. (18) and (19)).

As the temperature increases, the diffusivity increases and this makes the sensor respond faster (Fig.10b). In commercial instruments, thermistors are used for compensating temperature variation in sensor output (Fig.10a).

Cathode Diameter on Sensor Performance

Effect of Sensor Size on Flow Dependency: When a DO sensor is placed in a stagnant liquid a diffusion field is generated due to the consumption of oxygen at the cathode surface. The size of steady-state diffusion field is proportional to the size of the cathode. When the liquid is stirred, the diffusion gradient can no longer be extended beyond the liquid film around the membrane.

Since the diffusion gradient becomes steeper with decreasing liquid film thickness, the sensor output current increases with increase in liquid velocity as shown in Fig. 11. This flow dependency is larger for larger cathode area because the size of the stagnant diffusion field is proportionally larger. For proper operation of the sensor, the liquid has to be stirred beyond a certain level to maintain membrane control of oxygen diffusion.

The critical velocity, V_c, of the liquid is the velocity where the probe output reaches 95% (95-99% depending on the definition) of the steady state value. For a given liquid, V_c is smaller for smaller cathodes.

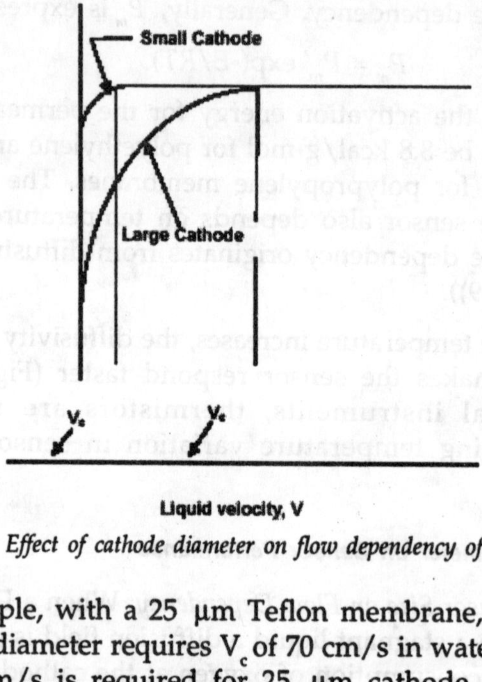

Figure 11. Effect of cathode diameter on flow dependency of DO sensor.

For example, with a 25 µm Teflon membrane, a cathode of 5 mm diameter requires V_c of 70 cm/s in water, whereas only 5 cm/s is required for 25 µm cathode. When the cathode diameter is less than 1 µm, the sensor becomes insensitive to liquid flow even without the membrane. In this case, the diffusion field of the cathode is so small that it is always contained inside the minimum liquid boundary layer around the cathode.

Effect on current output: One obvious effect of the cathode size is the area effect. Eq. (17)shows that the current output is directly proportional to the sensor area *A*. When the current produced by the sensor is too small (when the cathode area is small), the sensor signal tend to be more susceptible to noise. The current output should be greater than 10^{-6} to 10^{-7} range for ease of signal amplification.

Steam-Sterilizable Dissolved Oxygen Electrode

Fabrication of DO Sensor

DO Measurement: Oxygen must be supplied to growing microorganisms just like we need it for maintaining our lives. Oxygen dissolves very sparingly in water - about 8 ppm (parts per million by weight) at 20°C. Therefore, one of the major design variables for a bioreactor is the oxygen transfer rate (OTR) because the maximum growth rate of an aerobic microorganism depends on it.

The dissolved oxygen in bioreactors are usually measured by a dissolved oxygen (DO) sensor. In this Section, we will construct a steam-sterilizable DO sensor suitable for bioreactor applications. The idea is to appreciate the relationship between the sensor design/ fabrication methods and the sensor performance.

Materiuls of Construction : The following materials are needed:

— Electrode material

  — 75 mm (or 25 to 125 mm) diameter Pt wire, 3 cm length

  — a short length (20 cm) of Ag wire, 0.25 mm OD

  — 1 mil (1"x1") thickness Teflon membrane

  — 1M KCl solution in ethylene glycol

— Construction aids

  — lead glass capillary tube (2 mm OD, 0.5 mm ID), 15 cm long

  — glass tube (10 mm OD, 8 mm ID), 12 cm long

- a short length (20 cm) of copper wire, 0.25 mm OD
- a short length (1 cm) of 7 mm ID silicon tubing
- a short length (10 cm) of 2 mm ID rubber tubing (for use as a vent tube)
- a BNC female connector (obtainable from Radio Shack)
- epoxy (2 part, ribbon type)

— Chemicals for cleaning and plating

- 1M nitric acid solution
- Concentrated nitric/sulfuric (1:1) solution
- Methyl chloroform solution
- 0.1 M HCl solution

Fabrication: *Step 1.* Cut a short length of copper wire (20 cm) and attach 75 µm Pt wire by soldering

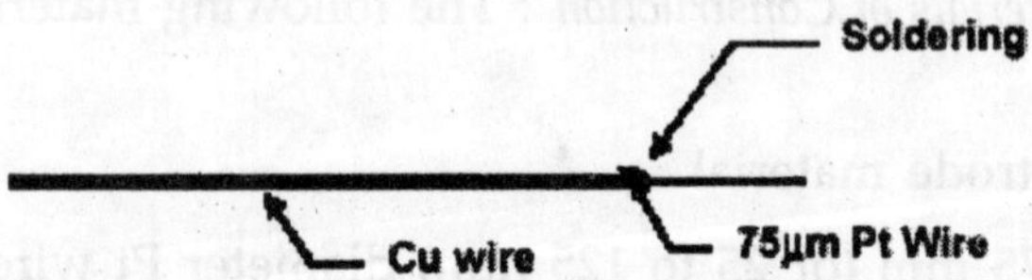

Step 2. Clean the Pt wire (just the Pt wire portion, not the Cu wire) by dipping in a 1:1 concentrated nitric and sulfuric acid solution. Wash with DI water.

Step 3. Place it in the glass capillary tube (cleaned) as shown. Flame the end until the glass melts and fuses around the Pt wire. A good wetting of the Pt wire with molten glass is necessary. Lead glass is best for this purpose because, it wets the metal well and the thermal expansion coefficient is closest to that of Pt.

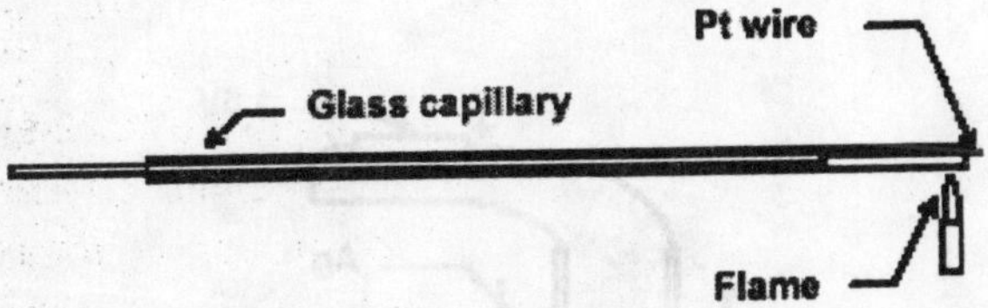

Step 4. Grind the flamed end flat (use a sand paper: start with a coarse one and then gradually use finer grit sand paper). The end should looks like:

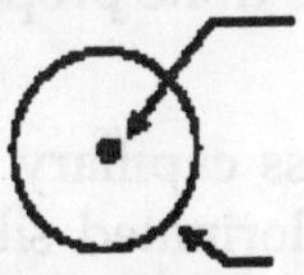

It is important that the glass seals the metal tightly around the metal wire. The goodness of the seal may be observed under a microscope but a better way is to measure the conductivity in an electrolyte solution (by using Ag/AgCl as the counter electrode). The conductivity should remain more or less constant. If it increases with time, the seal is not good and a DO sensor made with such a leaky seal tend to be unstable.

Step 5. Grind flat both ends of the 10 mm OD glass tube. Grind the edge of one end round.

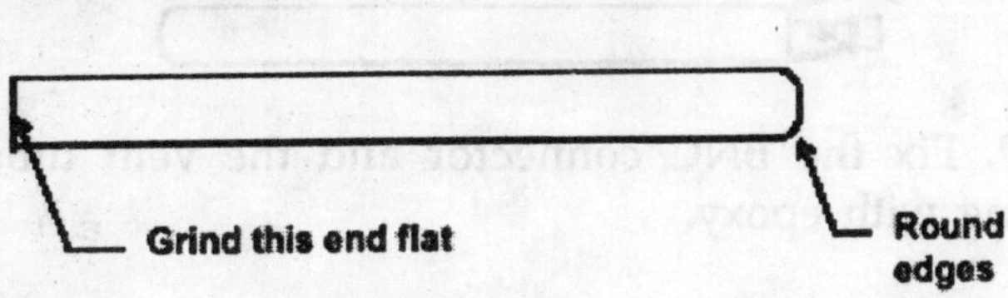

Step 6. Make a Ag/AgCl reference electrode by electrolytic chlorination of Ag wire.

— Clean Ag wire by dipping in 1 M nitric acid for 10 s.

— Anodize the wire in 0.1M HCl at current density of 0.4 mA/cm^2 for 30 min.

— Store in 1M KCl solution overnight before use.

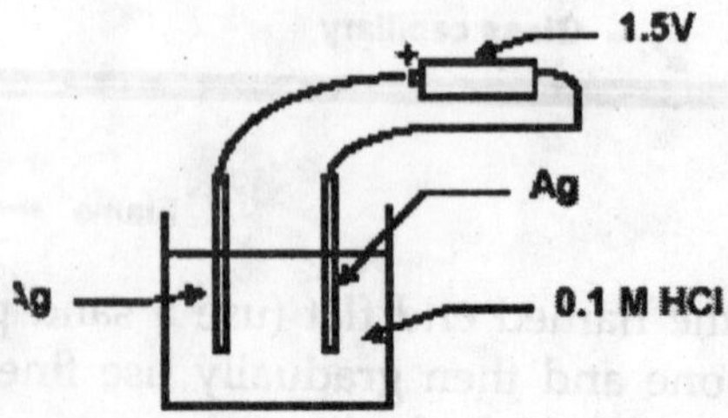

When the chlorination is done properly, the wire will have a brownish coat on it.

Step 7. Put the glass capillary tube in the large glass tube, and insert the chlorinated silver wire and the vent tube.

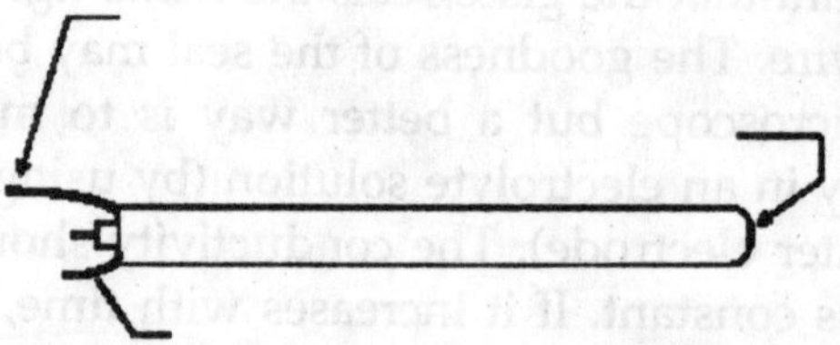

Step 8. Connect the copper wire to the center pin of BNC connector; connect the Ag/AgCl wire to the outer body of BNC by soldering.

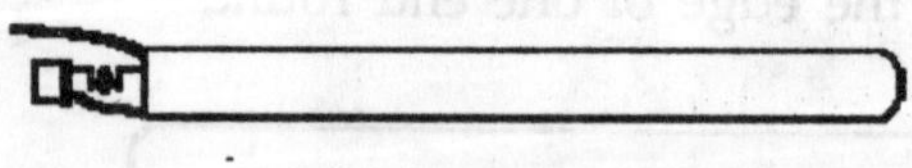

Step 9. Fix the BNC connector and the vent tube by molding with epoxy.

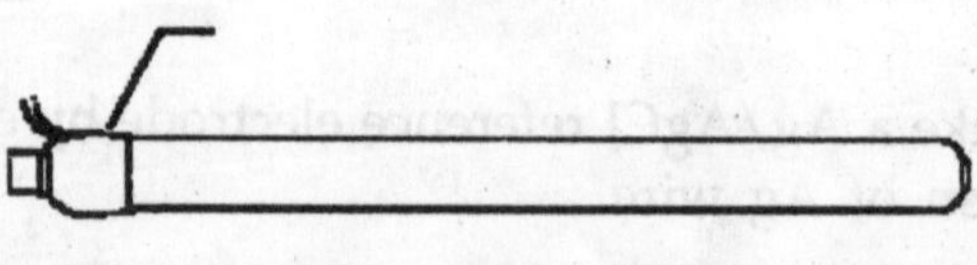

Step 10. Cut a ¾" length of silicone tubing and place it in methyl chloroform for 2 min. The tubing swells. Cover the free end of the sensor assembly with a Teflon

membrane and fix it tightly in place with the swelled silicone tubing. Make sure the membrane is tightly fit. Let the solvent evaporate.

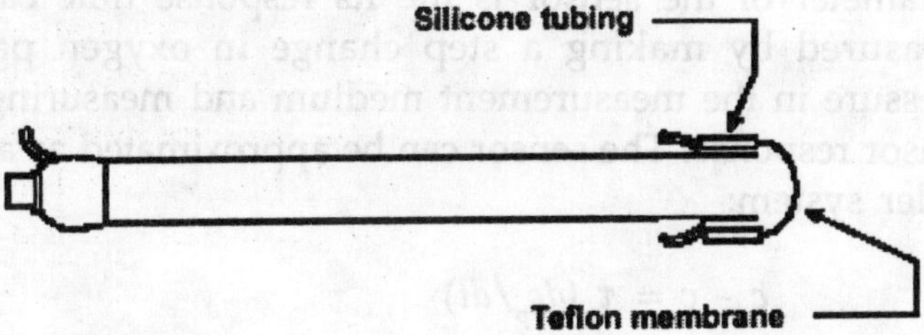

Step 11. Fill the electrolyte solution with a syringe via the vent hole. After the filling, clamp the vent tube.

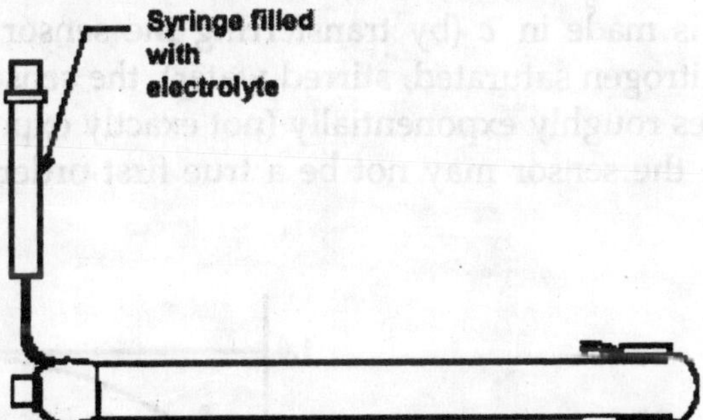

Calibration and k_La Measurement

Liquid Phase Calibration: Prepare an air saturated water by passing air bubbles into a small volume (100 mL) of water. Prepare a nitrogen saturated water in the same way. Connect the fabricated DO sensor to signal amplifying circuit, and then measure the voltage output for both water solutions. The liquids have to be agitated at high speed to obtain proper calibration. This is so-called a 'two point' calibration.

Gas Phase Calibration: Perform the calibration in gas phase by exposing the sensor to air. Do the same using nitrogen as the gas phase. Compare the two calibrations (between liquid and gas).

Measurement of Response: Time The other important parameter of the sensor is the Its response time can be measured by making a step change in oxygen partial pressure in the measurement medium and measuring the sensor response. The sensor can be approximated as a first order system:

$$c - c_p = \tau_p (dc_p/dt) \tag{22}$$

where c is the oxygen concentration in the measurement sample, c_p is the oxygen concentration measured by the sensor, and τ_p is the sensor time constant. When a step change is made in c (by transferring the sensor from air into a nitrogen saturated, stirred water), the sensor output decreases roughly exponentially (not exactly exponentially because the sensor may not be a true first order system).

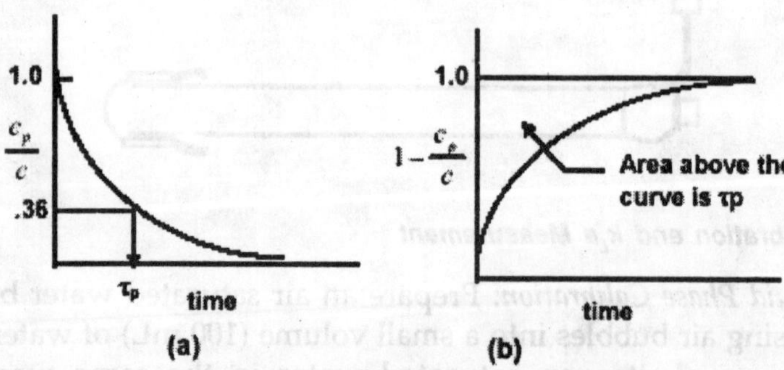

Figure 12. (a) Sensor response time measurement; (b) integral method for measuring the sensor time constant.

The time constant τ_p is the time when the sensor response reaches 63.7% of the ultimate response (Fig. 12a). The solution to Eq. (22) with the following boundary condition is an exponential function.

$$c = 1 \qquad \text{at } t = 0 \tag{23}$$

Note that a normalized concentration is used: *c* of 1 means 100% air saturation and 0 means nitrogen saturation. The solution is:

$$c/c_p = 1 - \exp(-t/\tau p) \tag{24}$$

Eq. (24) indicates that when t = τ_p, c/c_p will be 0.64. The time constant τ_p can also be determined conveniently by using an integral method - the area above the response curve is equal to τp (Fig. 12b). This method is especially useful when there is a lot of noise in the measured signal. The integration can be carried graphically using either trapezoidal rule of Simpson's rule.

Measurement of k_La: The oxygen absorption capability of a bioreactor is represented by k_La, the liquid phase overall volumetric mass transfer coefficient. DO sensor is used frequently to measure k_La. Typically, the reactor is first sparged with nitrogen and at time zero, the nitrogen is switched to air. The oxygen mass balance in the reactor yields:

$$dc/dt = kLa(c^* - c) \tag{25}$$

where c is the oxygen concentration in the reactor and c* is the oxygen concentration at the gas-liquid interface. This equation can be rearranged to:

$$c^* - c = \tau_k(dc/dt) \tag{26}$$

where

$$\tau_k = 1/k_La \tag{27}$$

Eqs. (22) and (26) can be solved simultaneously to obtain an expression for k_La. However, τ_k can be obtained graphically as shown in Fig. 13 when τ_p is known.

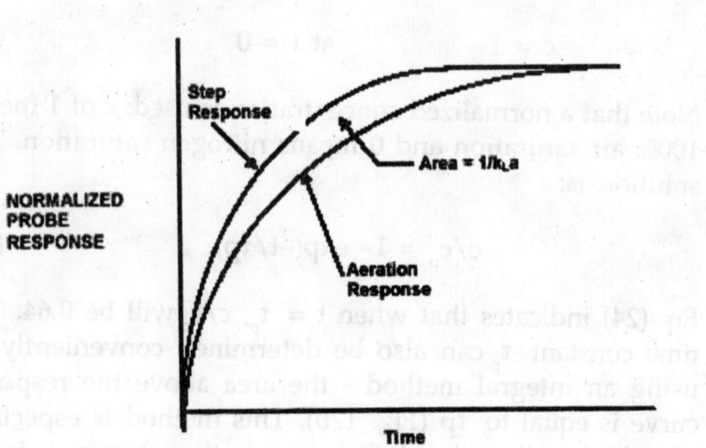

Figure 13. Measurement of kLa by integral method.

Caution in kLa Measurement: Note that the magnitude of τ_p depends on the liquid velocity in the vicinity of the sensor. Therefore, if a τ_p measured at one agitation rate is used for measuring k_La for different agitation rates, the results will be in error.

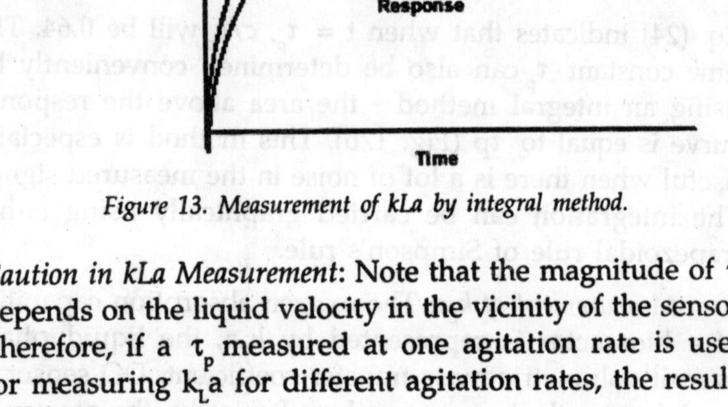

Figure 14. Signal amplifier circuit for H_2O_2 measurement.

A safe way is to used the same agitation rate for both τ_p and τ_k measurements. However, if τ_k is much greater than τ_p, such a precaution is not necessary.

Peroxide Sensor

Many oxidoreductase enzymes produce H_2O_2 as the reaction product. An example is glucose oxidase enzyme that produces H_2O_2 when glucose reacts with oxygen. H_2O_2 can be detected by polarography just as with DO sensor. The only difference is that the polarization voltage has to be + 0.7 V instead of - 0.7 V. The circuit given in Fig. 14 has to be used instead of the one given in Fig. 14(note that +0.7 V is applied using 78L05 voltage regulator).

Ion Selective Electrodes and Potentiometry

Definition of pH: Acids may be defined as proton donors and bases as proton acceptors:

$$HA = H^+ + A^- \tag{28}$$

(acid) (proton) (base)

The dissociation of weak acids and bases are equilibrium processes, and the equilibrium law can be applied to them:

$$K = \frac{[H^+][A^-]}{[HA]} \tag{29}$$

where K is the dissociation constant. For water,

$$H_2O \rightleftarrows H^+ + OH^-$$

and

$$K = \frac{[H^+][OH^-]}{[H_2O]} \tag{30}$$

K_w *for Water*: Since only a small fraction of water is dissociated, [H20] may be considered constant:

$$[H^+][OH^-] = K_w \tag{31}$$

Conductivity measurements have shown that Kw is 1x10-14 mol2/L2. 'The acidity or alkalinity of a solution can be measured by its hydrogen ion concentration but it is more convenient to use pH defined by:

$$pH = -\log[H+] \tag{32}$$

Therefore, from Eq. (31),

$$pH + pOH = 14 \tag{33}$$

When pH = pOH = 7, the solution is said to be in 'neutral' pH.

K_a *of Weak Acid:* From Eq. (29), pH of a weak acid can be described as:

$$pH = pK_a + \log([A^-]/[HA]) \tag{34}$$

Therefore, if K_a of an acid is known and pH is measured, the concentration of A^- ion can be calculated when the concentration of HA is known.

Measurement of pH

Glass type pH Electrode Usually, pH is measured by a glass electrode type pH sensor. The glass electrode consists of a glass membrane which is selectively permeable only to H+ and an internal Ag/AgCl electrode immersed in 0.1 M KCl solution (Fig. 15). When this glass electrode is immersed in a test solution of unknown pH and the voltage is measured against a standard electrode, the measured voltage follows Nernst equation:

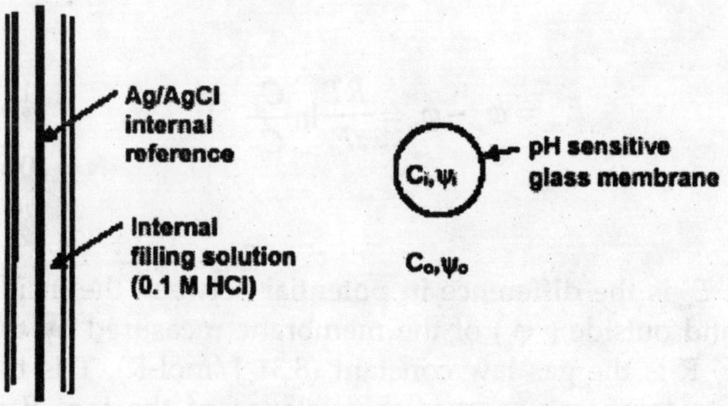

$$E_m = \varphi_o - \varphi_i = \frac{RT}{zF} \ln \frac{C_o}{C_i} \; (Nernst\ Eq.)$$

where R = 8.31 J/mole-K
F = 96,500 coul/mol
1 coul = 1 J/V
z = valence of ion

For H^+ at 20°C, 58.1 mV/decade of C^o

Figure 15. Glass membrane electrode for pH measurement.

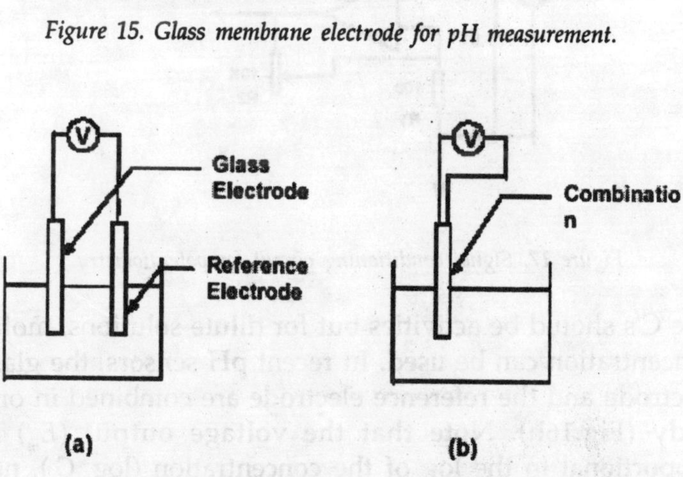

Figure 16. Measurement setup for pH: (a) separate glass electrode and reference electrode; (b) a combination electrode.

$$E_m = \varphi_o - \varphi_i = \frac{RT}{zF}\ln\frac{C_o}{C_i} \tag{35}$$

where E_m is the difference in potential between the inside (φ_i) and outside (φ_o) of the membrane measured by the sensor, R is the gas law constant (8.31 J/mol-K), T is the absolute temperature, z is the valance of the ion, F is Faraday constant (96,500 coul/mol; note that 1 coul = 1 J/V), and C_i and C_o are the concentrations of H^+ ions inside and outside of the membrane, respectively.

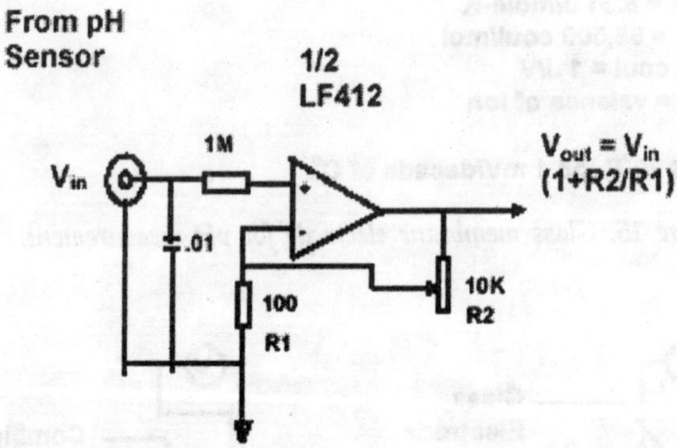

Figure 17. Signal conditioning circuit for potentiometry.

The C's should be activities but for dilute solutions, molar concentration can be used. In recent pH sensors, the glass electrode and the reference electrode are combined in one body (Fig.16b). Note that the voltage output (E_m) is proportional to the log of the concentration (log C_o), not the concentration itself (C_o) (Fig. 16).

Measurement circuit: From Eq. (35), a decade change in C_o will give E_m of 58.1 mV for H^+ at 20°C. This voltage output has to be amplified if one desires to take data, for example, with a computer. A circuit shown in Fig. 16 can be used for such an amplification. It uses one operational amplifier in a noninverting mode.

The operational amplifier has to be a FET-input type that has an input resistance on the order of 10^{12} ohms. This is because the resistance of the glass pH electrode is close to 5×10^9 ohms. To make voltage measurement from such a high resistance source, the input impedance of the amplifier has to be much greater. The other point is that only the non-inverting configuration (of the operational amplifier)

Other Ion-Selective Electrodes

pH measuring glass electrode is just one example of ion-selective electrodes (ISE). There are many different ISE's but it would be convenient to classify them in terms of the membrane used for partitioning.

Glass Membrane: These originate with the hydrogen-ion selective electrode having well-behaving glass membrane that has high mobility for H^+ ion. Subsequently, electrodes have been developed for other cations such as Na^+, K^+, and NH_4^+

Inorganic Salt Membrane: These electrodes are based on inorganic halides and sulphides, for example, silver salts, lanthanum fluoride and heavy-metal sulphides. These membranes have been produced from preparations ranging from whole crystals to dispersions in an inert matrix, such as polythene or silicon rubber. They are targeted at ions such as halides, CN^-, S^-, Ag^+, Cu^{++}, and Pb^{++}.

Organic Membrane: The electrodes based on neutral carriers generally have the highest selectivity in this class. However, cation exchangers or complexing agents or anion exchangers have been successfully employed in electrodes with liquid or solid membranes, selective to cations or anions, respectively.

Gas Sensing Electrode: These electrodes are an extension of ionselective measurements to detection of gaseous analytes. Gas-sensing probes are complete electrochemical cells, incorporating both the ion-selective electrode and a reference electrode within the sensor. Assay of the target gaseous sample is not performed directly, but is related to a changing parameter (usually pH) which can be monitored by an ion-selective electrode.

5

Bioreceptor Molecules

Enzymes

Biomolecular Ladder

Biomolecular Ladder Most of the chemical components of living organisms are organic compounds of carbon, many also containing oxygen and nitrogen. Although each living species contains various combination of these biomolecules, the diversity can be reduced to a few building blocks of common structure. It is possible to organize these building blocks on a hierarchical ladder, according to molecular weight (Figure 1).

The bottom of' the ladder is occupied by the low-molecular-weight gases (oxygen, nitrogen and carbon dioxide) and by water. These molecules together with the monoatomic ions, in particular Na^+, K^+, Mg^{2+}, Ca^{2+}, Cl^- and the elements P, S, Mn, Fe, Co, Cu and Zn are most common in further involvement as progress is made up each rung of the molecular-weight ladder.

The monitoring of these ions is frequently used to follow the metabolic state of a patient under care and is commonly achieved using ion selective electrode sensors, where the recognition surface of the sensor is provided by a membrane, containing a molecule with a selectivity for the target ion. The oxygen and carbon dioxide

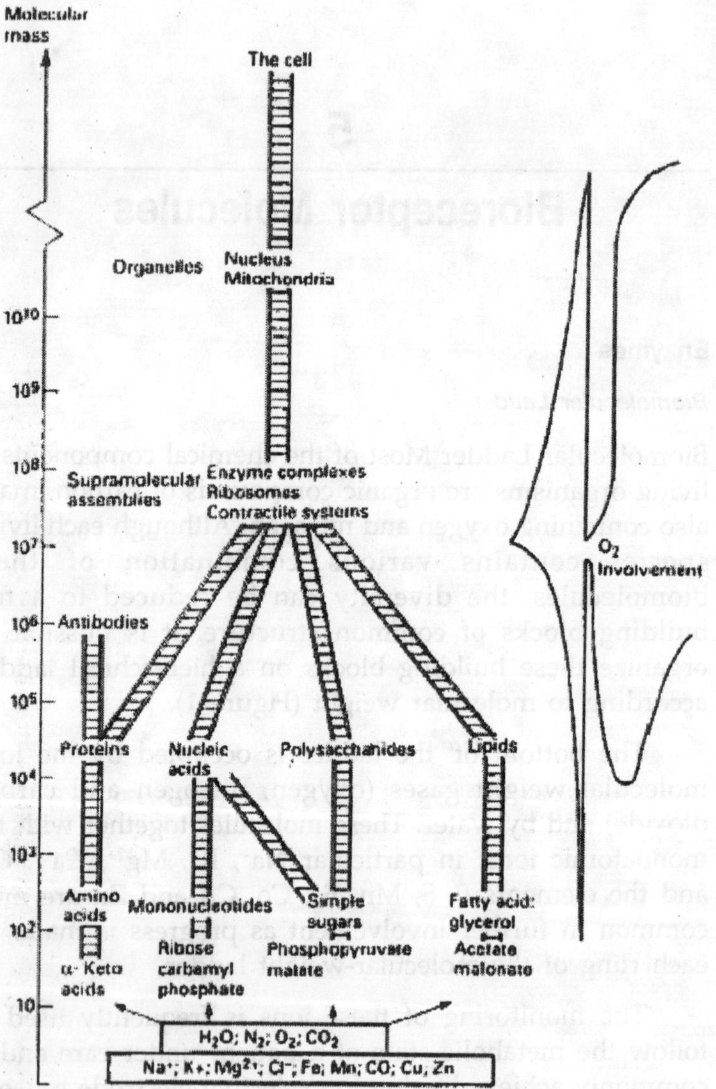

Figure 1. Biomolecular ladder

electrodes form two of the wellestablished base sensors to which analyte specific reactions have been linked via a biorecognition macromolecule. Note where 'protein', 'antibody', and 'enzyme complex' lie in the ladder. These are the major molecules with molecular-recognition capabilities that are utilized in biosensor development.

The food that we eat consists mainly of proteins, carboydrates (or polysaccharides), and lipids. These nutrients are broken down in our body with the aid of oxygen that we breathe. This breakdown process called 'catabolism' produces the basic building blocks of the body - amino acids, simple sugars, and fatty acids. Also, energy is produced (and stored in the form of ATP) during the catabolism which is essentially a partial oxidation process.

These building blocks and energy are then used to make the main components of our body - proteins, polysaccharides, and lipids. These components are used to make supramolecular assemblics such as enzyme complexes, ribosomes (this is where proteins are made), etc. The synthesis processes are collectively called the 'anabolism' which is dictated by the genetic codes written on the genes of the cell (the DNA helix).

Protein Structure Proteins are polymers of amino acids. All amino acids have a common structure - a carbon molecule with a carboxyl group (–COOH), an amino group (–NH_2), and R group (Fig. 2a). It is the R group which makes each amino acid unique. For example when R = CH_3, it is called alanine. The amino acids are connected together by a peptide bond, which is the bond between -COOH of one amino acid and -NH_2 of the other (Fig. 2b).

The proteins are made in the riobosomes (an organel made of ribosomal RNAs) of the cells. The strucutre and the function of the proteins depend not only on the amino acicd sequence (called the primary structure), but also on the conformation.

(a) Basic structure of amino acid

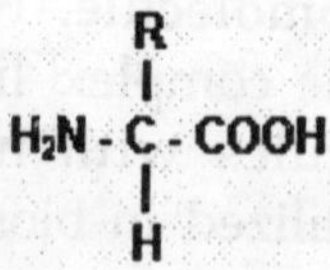

(b) Formation of peptide bond

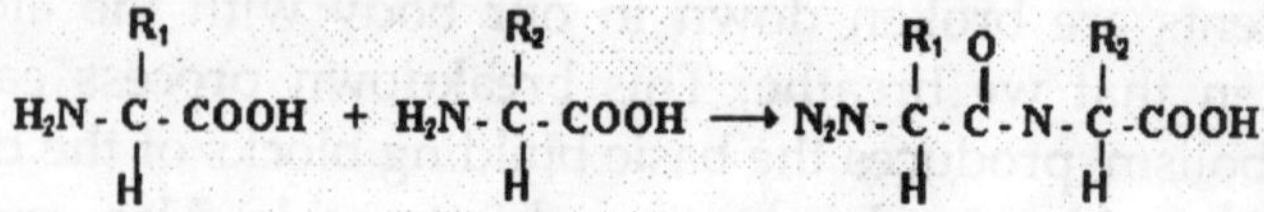

(c) Limited rotation in peptide chain

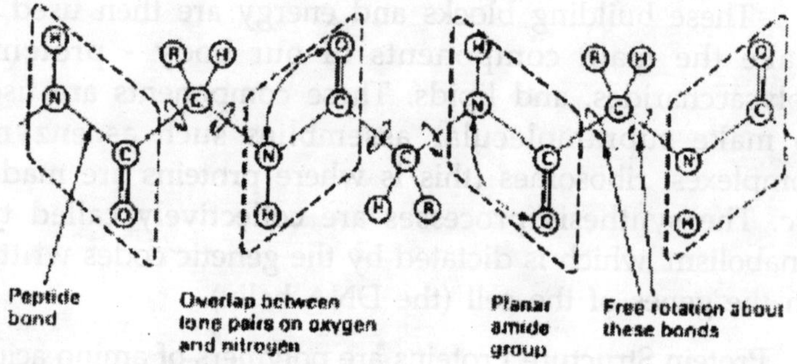

(d) non-covalent inter- and intra-molecular bonds in peptide strands.

Figure 2. (a) Basic structure of amino acid; (b) formation of peptide bond; (c) limited rotation in the peptide strand; (d) non-covalent inter- and intramolecular bonds in peptide strands.

The peptide bond and the disulphide bridges impart certain restrictions on the structure (Fig. 2c). The peptide strands are further organized by interactions between residue side chains. The nature of the bonds include hydrogen bond, ionic bond, and hydrophobic bond (Fig. 2d).

Ionic property of proteins

The net result of all the interactions between the amino acids is that there is a spontaneous folding of a protein to give a unique structure. All the amillo acids have at least two groups capable of existing in ionic form. The α-(carboxyl groul),—COOH, can lose H^+ to become COO^-. The reaction is pH dependent, and is characterized by a pK_a typically in the range 2 to 3.

Similarly the α-amino group, NH_2, can be protonated to give NH_3^+ and has a pK_a value of about 10. Therefore, between about pH 4 and 9 the amino acid exists as a dipolar ion zwitterion with little net charge. At the isoelectric point, pI, the protein has no net charge, and it will not move in an electric field.

Ioselectrical Point

Where R contains no ionizable groups,

$$pI = \frac{(pK_a^{NH2} + pK_a^{COOH})}{2} \quad (1)$$

The movement of the amino acids under the influence of an electric field allows their separation and identification; it can be used as a powerful assay technique. The dual polarity feature accounts for many of the properties of amino acids, e.g. the large dipole moments, the high solubility in water and low solubility in organic solvents.

Proteins' ionic behavior

By analogy, it would be expected that each peptide strand would also contain at least two ionizable groups, but since the α-carboxyl or α-amino groups are now involved in peptide-bond formation, they are not available for ionization. The zwitterionic behaviour is therefore more restricted to the terminal amino group and the terminal carboxyl group. These groups are considerably further away from one another than they would be in free amino acids and so the electrostatic interactions between them are diminished and their pK_a values are lower than in the α-amino acid.

It follows that the groups in proteins that are principally involved in acid-base equilibria are the side-chain R groups. Ordinarily there may be 50-60 titratable groups per 100,000 molecular weight of protein. Therefore, the titration curves for proteins are complex and difticult to interpret.

Molecular recognition by enzyme

The biorecognition properties of protein molecules will depend almost entirely on the amino acids of the exposed surfaces. Weak non-covalent interactions can occur between the residues on the exposed surfaces of the protein and other non-protein molecules. If a sufficient number of these weak bonds are formed simultaneously with the incoming molecule, then the molecule can bind tightly to the protein. Obviously for this to occur the molecule must fit precisely into the binding site on the protein surface. This feature is analogous to the recognition surface of the model biosensor.

Enzyme complexes

One of the most important functions of proteins is to act as catalysts or enzymes for chemical reactions. These enzymes are able to stabilize the transition state between a substrate and its products by interactions at the binding site (of the substrate). The activation of most biochemical reactions fall in the range of 40-80 kcal/mol without the enzyme. The enzymes lower this activation energy. For example, the splitting of H_2O_2 takes 75.4 kJ/mol without the enzyme, whereas a catalase enzyme lowers the activation energy to 23 kJ/mol.

Substrate specificity by the enzyme is provided by the surface interactions and this characterisitcs is exploited in the development of enzylnebased biosellsors. The none-covalent binding of the enzyme substrate transition state lowers the activation energy for the reaction and thus catalyzes the reaction.

Classification of Enzyme

There are approximately 3000 known enzymes. These enzymes are classified into six categories based on the type of reaction they catalyze.

1. *Oxido- reductase*: Oxidizes or reduces by transfer of hydrogen or electrons.

   (a) dehydrogenases:

   $SH_2 + A \leftrightarrow S + AH_2$ (S: Substrate, A: acceptor)

   Example:

   Lactate dehydrogenase: L-lactate + NAD $\leftrightarrow$ pryuvate + NADH + H^+

   (b) oxidases:

   $SH_2 + 1/2\ O_2 \rightarrow S + H_2O$ or

   $SH_2 + O_2 \rightarrow S + H_2O_2$

   Eample

Glucose oxidase: β-D-glucose + O_2 → gluconolactone + H_2O_2

(c) peroxidases:

$2SH + H_2O_2 \rightarrow 2S + 2H_2O$ or

$2S + 2H+ + H_2O_2 \rightarrow 2S+ + 2H_2O$

Example

Horseradish peroxidase:

$2[Fe(CN)_6]_4- + 2H^+ \rightarrow 2[Fe(CN)_6]_3 + 2H_2O$

(d) oxygenases:

$SH + DH + O_2 \rightarrow S\text{-}OH + D + H_2O$

Example

Lactate 2-monooxygenase:

L-lactate + O_2 → acetate + CO_2 + H_2O

2. *Transferase*: Transfers C-, N-, P-, or S-containing functional groups such as aldehydes and ketones, glycosils, acyls, phosphates, and sulfur containing groups.

   $AX + B \leftrightarrow A + BX$

   Example

   Hexokinase: D-hexose + ATP ↔ D-hexose-6-phosphate + ADP

3. *Hydrolase*: Hydrolyses esters, anhydrides, peptide bonds, other C-N bonds, glycosides

   Example

   Cholesterol esterase:

   cholesterol ester + H_2O → cholesterol + fatty acid

   Glucoamylase: amylose + $n\ H_2O \rightarrow n\ \beta$-*D*-glucose

4. *Lyase*: Adds to double bonds:

$$> C = C <$$

$$> C = O$$

$$> C = N$$

5. *Isomerase*: Isomerizes optical iomers

   Example

   Glucose iomerase: *D*-glucose ↔ *D*-fructose

6. Ligase: Splits C-C, C-O, C-N, C-S and C-halogen bonds without hydrolysis or oxidation, mostly with ATP

   Example

   Pyruvate carboxylase:

   pyruvate + HCO_3^- + ATP ↔ oxaloacetate + ADP + P_i

Coenzymes, Prosthetic group, Effectors

Sometimes the surface cavity does not act as a catalytic site until it is modified by a second incoming molecule. These participants known as the coenzymes are non-peptide molecules capable of completing the binding site for the transition state. Other molecules that do the similar function are prothetic group, and effectors.

Coenzyme: Coenzyme is a non-peptide molecule capable of completing the binding site for the transition state. Examples include many vitamin derivative such as coenzyme A, thiamine, pyrophosphate, vitamin B12

Prosthetic Group: Prosthetic group is the same as the coenzyme but are tightly bound to the enzyme. When they are split off, the enzyme is mostly denatured. Examples include flavin nucleotides and heme.

Effectors: Effectors accelerate (activators) or block (inhibitors) enzyme reaction Examples of activators include

Mg^{++}, Ca^{++}, Zn^{++}, K^{+}, and Na^{+}, while the examples for the inhibitors include Hg, and substrate analogs.Table 3.1. lists functions of some of the important coenzymes and prostshetic groups.

Table 1. Function of some important coenzymes and prosthetic groups.

Compound	Function
(a) Oxidoreduction	
Nicotinamide adenine dinucleotide (NAD)	hydrogen transfer
Nicotinamide adenine dinucleotide phosphate (NADP)	hydrogen transfer
Flavin monocucleotide	hydrogen transfer
Flavin adenine dinucleotide (FAD)	electron transfer
Heme (cytochromes)	
Ferredoxins	electron transfer
(b) Group Transfer	
Pyridoxalphosphate	transamination,
decarboxylation	
Adenosine triphosphate	phosphate group donor
Tetrahydrofolic acid	C1 group transfer
Biotin	carboxylation, decarboxylation
Coenzyme A	transacylation
Thiamine pyrophosphate (Vitamin B1)	C2-group transfer
Riboflavin	hydrogen transfer
5′-deoxyadenosyl-cobalamine	transfer of methyl group, isomerization

Enzyme Kinetics

Enzyme-Catalyzed Reactions

Reaction Velocity: The time course of an enzymatic reaction permits one to deduce the substrate affinity, the catalytic mechanism in the active center, and the efficiency of the enzyme (maximum rate, turnover number). The rate of an enzyme-catalyzed single reactant reaction depends on the concentration of substrate and product, respectively. The velocity of the reaction V is:

$$V = - dS/dt = dP/dt \quad (2)$$

where first term is the rate of dissapearance of substrate *S* and second term is the rate of appearance of product *P* (both *S* and *P* are in concentration).

Behavior of Initial Rates: The initial rate (V_o) is determined by extrapolating the slope of the time course of the substrate or product concentration to time zero (Fig. 3). The dependence of V_o on the substrate concentration, S (at constant enzyme concentration), is shown in Fig. 4. It reflects the typical substrate saturation. At first, V_o increases proportionally to the amount of substrate.

Upon further enhancement of substrate concentration V_o levels off. The curve asymptotically approaches a maximum value, Vmax. When this plateau is reached, a change of *S* does not lead to a measurable change of V_o: the enzyme is saturated by substrate and has thus reached the limit of its efficiency.

Micahaelis- Menten Kinetics : These kinetics result from the fast and reversible formation of an enzyme-substrate complex, ES, which dissociates in a second, slower reaction under liberation of the product, P (Fig. 5):

$$E + S \quad ES \quad E + P \quad (3)$$

Because the second reaction is rate-limiting, at very high substrate concentration almost all enzyme is present as enzyme-substrate com-plex. Under these conditions a steady state is reached in which the enzyme is steadily saturated by substrate and the initial rate is at a maximum (V_{max}). This relation between substrate concentration and reaction rate may be described by the Michaelis- Menten equation:

$$V_o = \frac{V_{max} S}{K_M + S} \tag{4}$$

where K_M is the Michaelis constant of the enzyme for the given sub-strate. K_M may also be described by:

$$K_M = \frac{k_{-1} + k_{+2}}{k_{+1}} \tag{5}$$

Meaning of K_M: The relevance of K_M becomes evident at $S = K_M$. Then $V_o = V_{max}/2$, i.e., K_M is the substrate concentration at which the reaction rate is half maximum (Fig. 4). The K_M value characterizes the affinity between the substrate and the enzyme. At known K_M and V_{max}, V_o can be calculated for each value of substrate concentration. A low K_M value reflects high affinity. At substrate concentrations $S \ll K_M$, the reaction rate is directly proportional to the substrate concentration (first order reaction); at high substrate concentration ($S \gg K_M$) the reaction is zero order and is no longer dependent on the substrate concentration but only on the enzyme activity.

Line-Weaver- Burk Plot: To calculate K_M and V_{max} (and inhibitor constants) it is advantageous to transform the Michaelis-Menten relation so as to obtain linear relationships between S and V_o that can be evaluated graphically. An example is the Line-Weaver-Burk equation, containing the reciprocal values of V_o and S:

$$\frac{1}{V_o} \frac{1}{V_{max}}\left(1+\frac{K_M}{S}\right) \tag{6}$$

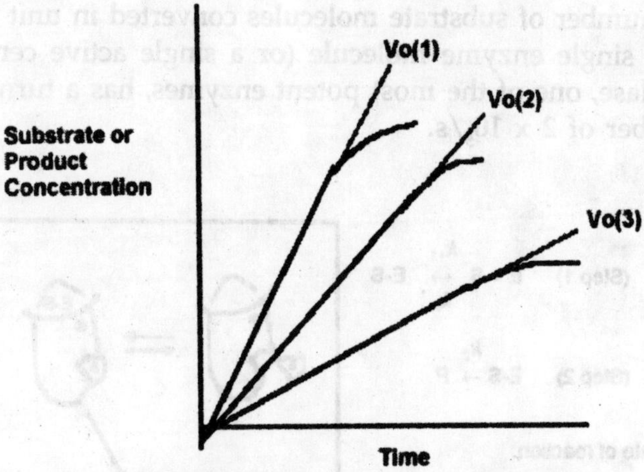

Figure 3. Determination of initial rates at different substrate concentrations.

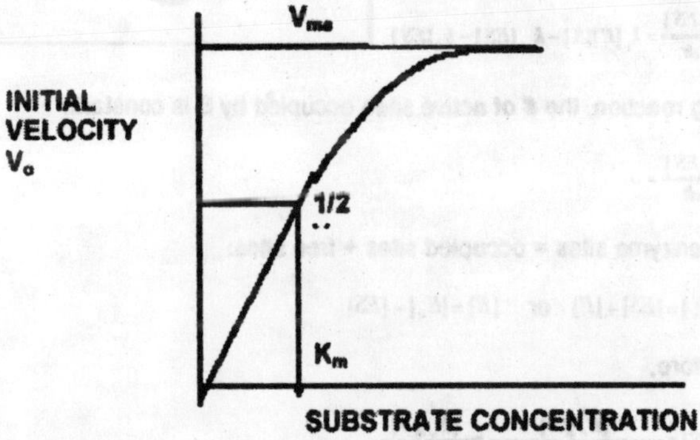

Figure 4. A plot of V_o vs. substrate concentration S.

Enzyme Activity and Enzyme Concentration

Turnover Number: In addition to K_M, and V_{max}, the turnover number (molar activity) and the specific activity are important parameters for the characterization of enzyme reactions. Both are determined under substrate saturation. With highly purified enzymes the turnover number reflects the number of substrate molecules converted in unit time by a single enzyme molecule (or a single active center). Catalase, one of the most potent enzymes, has a turnover number of 2 x 10_5/s.

(Step 1) $\quad E + S \underset{k_{-1}}{\overset{k_{+1}}{\leftrightarrow}} E\text{-}S$

(Step 2) $\quad E\text{-}S \xrightarrow{k_2} P$

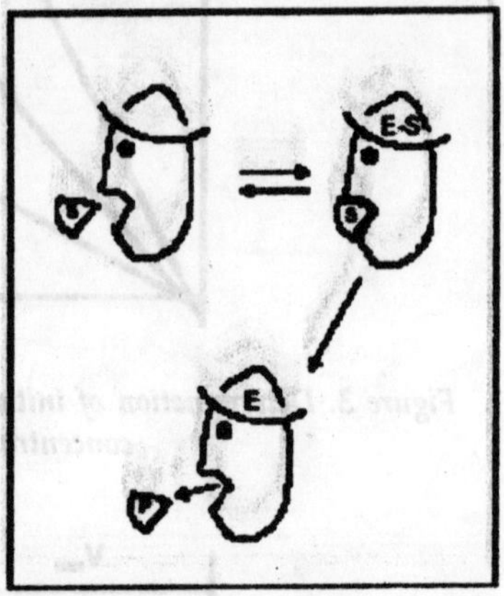

Rate of reaction:

$$V = -\frac{d[S]}{dt} = \frac{d[P]}{dt} = k_2[ES]$$

Net formation rate of ES:

$$\frac{d[ES]}{dt} = k_1[E][S] - k_{-1}[ES] - k_2[ES]$$

During reaction, the # of active sites occupied by S is constant:

$$\frac{d[ES]}{dt} = 0$$

Total enzyme sites = occupied sites + free sites:

$$[E_o] = [ES] + [E] \quad \text{or} \quad [E] = [E_o] - [ES]$$

Therefore,

$$[ES] = \frac{[E_o][S]}{(k_{-1} + k_2)/k_1 + [S]} = \frac{[E_o][S]}{K_M + [S]}$$

$$\boxed{V = -\frac{d[S]}{dt} = k_2[ES] = \frac{k_2[E_o][S]}{K_M + [S]} = \frac{V_{max}[S]}{K_M + [S]}}$$

Figure 5. Enzyme Kinetics

Specific Activity: The specific activity of enzymes is given in units. One international unit (IU) is the amount of enzyme consuming or forming 1 µmol substrate or 1 ?mol product per minute under standard conditions. The base unit is 1 katal, corresponding to the amount of enzyme converting 1 mol substrate per second:

$$1 \text{ kat} = 6\text{x}10^7 \text{ IU},$$

$$1 \text{ IU} = 16.67 \text{ nkat}.$$

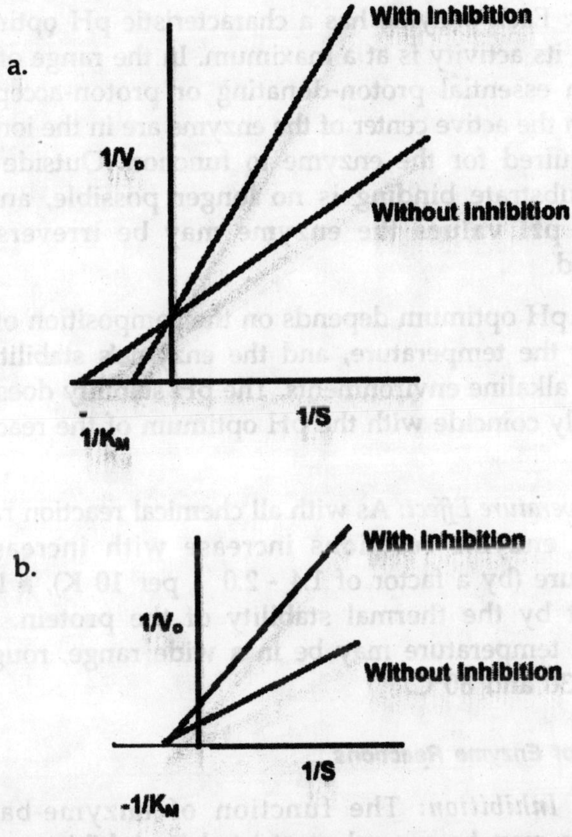

Figure 6. (a) Competitive inhibition; (b) non-competitive inhibition

Usually U is used instead of IU. For the quantitative determination of enzyme activity, initial rates are measured at different enzyme concentrations and near substrate saturation, in a suitable temperature range (25-

37°C) and at optimal pH. In a certain range the enzyme activity is proportional to the enzyme concentration. The enzyme activity of a sample can be estimated from the linear part of the plot.

pH and Temperature Dependence

pH Effect: Each enzyme has a characteristic pH optimum at which its activity is at a maximum. In the range of this optimum essential proton-donating or proton-accepting groups in the active center of the enzyme are in the ionized state required for the enzyme to function. Outside this range, substrate binding is no longer possible, and at extreme pH values the enzyme may be irreversibly denatured.

The pH optimum depends on the composition of the medium, the temperature, and the enzyme's stability in acid and alkaline environments. The pH stability does not necessarily coincide with the pH optimum of the reaction rate.

Temperature Effect: As with all chemical reaction rates, those of enzyme reactions increase with increasing temperature (by a factor of 1.4 - 2.0 per 10 K), a limit being set by the thermal stability of the protein. The optimum temperature may be in a wide range, roughly between 30 and 80°C.

Inhibition of Enzyme Reactions

Types of Inhibition: The function of enzyme-based biosensors may be severely restricted by inhibitors. The inhibition is either reversible or results in an irre-versible inactivation of the enzyme.

Competitive Inhibition: Inhibitors structurally related to the substrate may be bound to the enzyme active center and compete with the substrate (competitive inhibition).

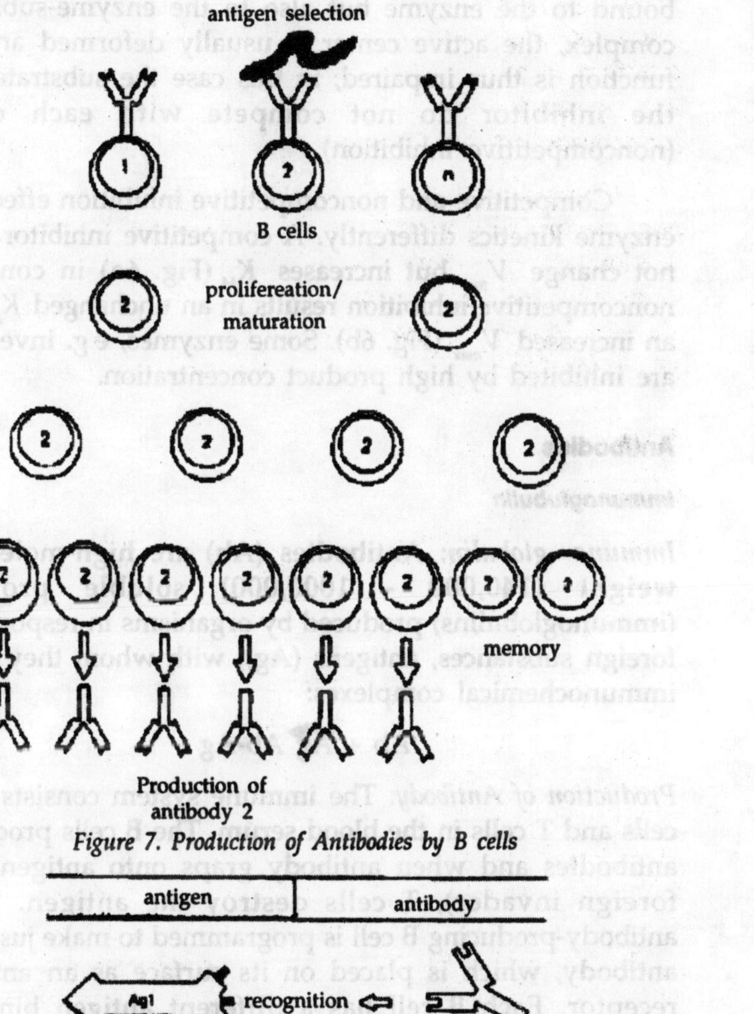

Figure 7. Production of Antibodies by B cells

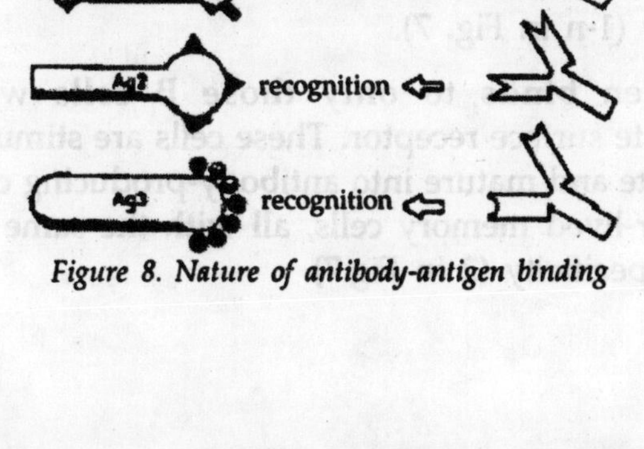

Figure 8. Nature of antibody-antigen binding

Non- competitive Inhibition: If the inhibitor is not only bound to the enzyme but also to the enzyme-substrate complex, the active center is usually deformed and its function is thus impaired; in this case the substrate and the inhibitor do not compete with each other (noncompetitive inhibition).

Competitive and noncompetitive inhibition effect the enzyme kinetics differently. A competitive inhibitor does not change V_{max} but increases K_M (Fig. 6a) in contrast, noncompetitive inhibition results in an unchanged K_M and an increased V_{max} (Fig. 6b). Some enzymes, e.g. invertase, are inhibited by high product concentration.

Antibodies

Immunoglubulin

Immuno- globulin: Antibodies (Ab) are high-molecular weight (140,000 - 1000,000) soluble proteins (immunoglobulins) produced by organisms in response to foreign substances, antigens (Ag), with whom they form immunochemical complexes:

$$Ab + Ag \rightarrow Ab\text{-}Ag$$

Production of Antibody: The immune system consists of B cells and T cells in the blood serum. The B cells produces antibodies and when antibody graps onto antigen (the foreign invader), T cells destroy the antigen. Each antibody-producing B cell is programmed to make just one antibody, which is placed on its surface as an antigen receptor. Each B cell has a different antigen binding specificity (1-n in Fig. 7).

Antigen binds to only those B cells with the appropriate surface receptor. These cells are stimulated to prolifereate and mature into antibody-producing cells and the longer-lived memory cells, all with the same antigen binding specificity (2 in Fig.7).

Structure of Antibody: The basic structure of all immunoglobulin molecules is a unit consisting of two identical light polypeptide chains and two identical heavy polypeptide chains linked together by disulphide bonds. The class and subclass of an immunoglobulin molecule is determined by its heavy chain type.

Thus the four human IgG subclasses (IgG1, IgG2, IgG3 and IgG4) have heavy chains called y1, y2, y3, and y4 which differ only slightly although all are recognizably y heavy chains. The differences between the various subclasses within an immunoglobulin class are less than the differences between the different classes; thus IgG1 is more closely related to IgG2, 3, or 4 than to IgA, IgM, IgD or IgE.

Subclasses of Antibodies: The four subclasses of human IgG occur in the approxi-mate proportions of 66, 23, 7 and 4 per cent respectively. There are also known to be subclasses of human IgA (IgA1 and IgA2) but none have been unambiguously described for the other classes. Immunoglobulin subclasses appear to have arisen after speciation and the human subclasses cannot be compared with, for example, the four known subclasses of IgG which have been identified in the mouse.

Properties of Immunoglobulins

General: All immunoglobulins appear to be glycoproteins but the carbohydrate content ranges from 2-3% for IgG to 12-14% for IgM, IgD and IgE. Each class possesses a characteristic type of heavy chain. Thus IgG possesses γ chains; IgM, μ chains; IgA, α chains, igD, δ chains and IgE, ε chains. Variation in heavy chain structure within a class gives rise to immunoglobulin subclasses.

For example the human IgG pool consists of four subclasses reflecting four distinct types of γ heavy chain. The physicochemical properties of the immunoglobulins vary between the different classes. Note that IgA occurs in a dimeric form {sigA) in association with a protein chain termed the secretory piece. The diversity of structure of the different classes suggests that they perform different functions, in addition to their primary function of antigen binding. In spite of this diver-sity all antibodies have a common basic structure.

IgG: IgG is the major immunoglobulin in normal human serum accounting for 70-75% of the total immunoglobulin pool. IgG is a monomeric protein with a sedimentation coefficient of 7S and a molecular weight of 146,000. However, studies of IgG subclasses have indicated that IgG3 proteins are slightly larger than the other subclasses and this increase is due to the slightly heavier y3 chain. The IgG class is distributed evenly between the intra- and extravascular pools, is the major antibody of secondary immune responses and the exclusive anti-toxin class.

IgM: IgM accounts for about 10% of the immunoglobulin pool. The molecule has a pentameric structure in which individual heavy chains have a molecular weight of approximately 65,000 and the whole molecule has a molecular weight of 970,000. This protein is largely confined to the intravascular pool and is the predominant 'early' antibody frequently directed against antig,enically complex infectious organisms.

IgA: IgA represents 15-20% of the human serum immuno-globulin pool . In man more than 80% of IgA occurs as the basic four chain monomer but in most mammals the IgA in serum is mainly polymeric, occurring mostly as a dimer. IgA is the predominant immunoglobulin in sero-mucous secretions such as saliva, tracheobronchial secretions, colostrum, milk and genitourinary secretions.

Secretory IgA (sigA) which may be of either subclass, exists mainly in the 1 15, dimeric form and has a molecular weight of 385,000. sigA is abundant in seromucous secretions and is protected from proteolysis by combination with another protein - the secretory component.

IgD: IgD accounts for less than 1% of the total plasma immunoglobulin but it is known to be present in large quantities on the membrane of many circulating B Iymphocytes. The precise biological function of this class is not known but it may play a role-in antigen-triggered Iymphocyte differentiation.

IgE: IgE though a trace serum protein, is found on the surface membrane of basophils and mast cells in all individuals. This class may play a role in active immunity to helminthic parasites but in Western countries is more commonly associated with immediate hypersensitivity diseases such as asthma and hayfever.

Interaction of Antigen-Antibody

Lock and Key Type Binding: Foreign molecules that generate antibodies are called antigens. Antigen molecules each have a set of antigenic determinants also called epitopes (Fig. 8). The epitopes on one antigen are usually different from those on another. Some antigens have repeated epitopes. Epitopes are molecular shapes recognized by the antibodies and cells of the adaptive immune system. Each cell recognizes one epitope rather than the whole antigen. Even simple microorganisms have many different antigens.

Nature of Binding Forces: There are several types of intermolecular attractive forces binding antigen to antibody. These forces require the close approach of the interacting groups. Hydrogen bonding results from the formation of hydrogen bridges between appropriate atoms; electrostatic forces are due to the attraction of oppositely charged groups located on two protein side chains.

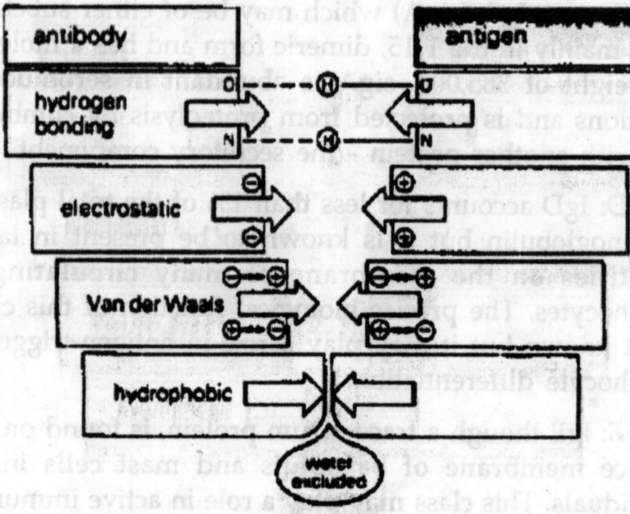

Figure 9. Types of intermolecular attractive forces binding antigen to antibody

Van der Waals bonds are generated by the interaction between electron clouds (here represented as induced oscillating dipoles) and hydrophobic bonds (which may contribute up to half the total strength of the antigen-antibody bond) rely upon the association of non-polar, hydrophobic groups so that contact with water molecules is minimized. The distance of separation between the interacting groups which produces optimum binding varies for the different types of bond (Fig. 9).

Antibody Affinity

Nature of Affinity: The affinity with which antibody binds antigen results from a balance between the attractive and repulsive forces. A high affinity antibody impies a good fit and conversely, a low affinity antibody implies a poor fit (Fig. 10).

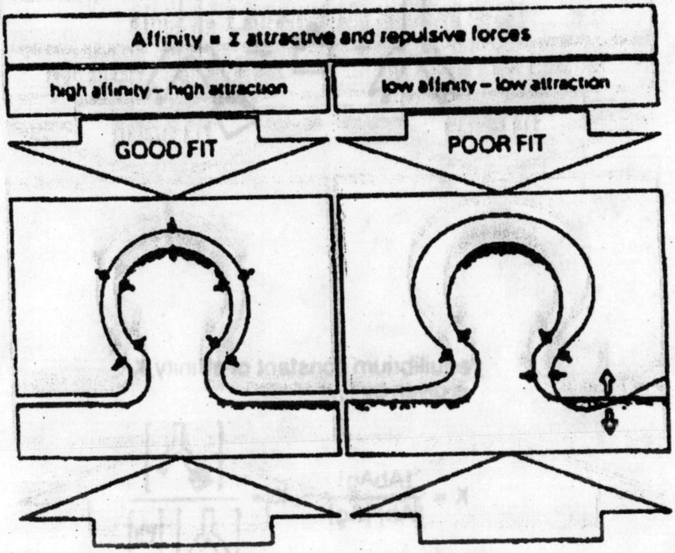

Figure 10. Antibody Affinity

Affinity Constant, K_a: All antigen-antibody reactions are reversible and the equilibrium constant K is called, the affinity constant (Fig. 11):

$$Ab + Ag \rightleftarrows Ab\text{-}Ag$$

$$K_a = \frac{[Ag - Ab]}{[Ag][Ab]}$$

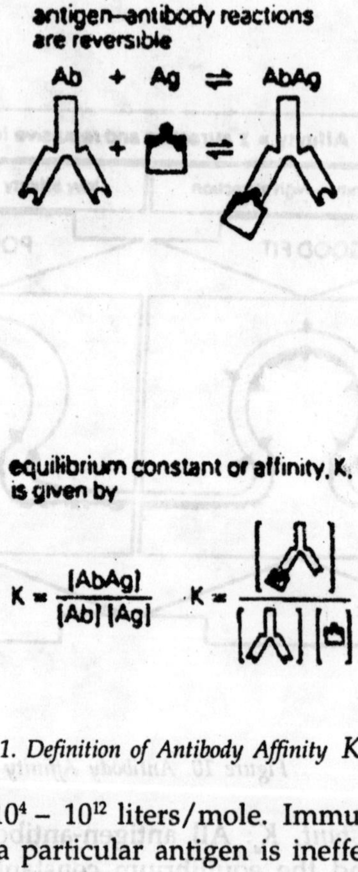

Figure 11. Definition of Antibody Affinity K_a

K_a ranges from $10^4 - 10^{12}$ liters/mole. Immunoglobulins with $K_a < 10^4$ for a particular antigen is ineffective.

Experimental Determination of K_a

Suppose we start with an antibody concentration of $[Ab]_o$ and vary antigen concentration $[Ag]$. At equilibrium, the antigen concentration will be:

$$[Ab]_e = [Ab]_o - [AbAg]_e \tag{8}$$

Let Kd be reciprocal of Ka:

$$K_d = \frac{1}{K_a} = \frac{[Ab]_e[Ag]_e}{[AbAg]_e} \tag{9}$$

Substituting Eq. (8) in Eq. (9):

$$[AbAg]_e = \left(\frac{[Ag]_e}{K_d + [Ag]_e}\right)[Ab]_o \tag{10}$$

In reciprocal form,

$$\frac{1}{[AbAg]_e} = \frac{1}{[Ab]_0} + \left(\frac{K_d}{[Ab]_o}\right)\frac{1}{[Ag]_e} \tag{11}$$

Thereforc, from a plot of $1/[AbAg]_e$ vs. $1/[Ag]_e$, K_d can be obtained experimentally, and $K_a = 1/K_d$. This process is shown graphically in Fig. 12b. K_d is equal to the reciprocal concentration of free antigen necessary to occupy half of the antigen binding sites of the antibody (Fig. 12a).

Avidity Multivalent binding between antibody and antigen (avidity or functional affinity) results in a considerable increase in stability as measured by the equilibrium constant, compared to simple monovalent binding.

This is sometimes referred to as the 'bonus effect' of multivalency. Thus there may be a 10^3 fold increase in binding energy of IgG when both valencies (combining sites) are utilized, and 10^7 fold increase when IgM binds antigen in a multivalent manner. Monovalent antigen combines with multivalent antibody with no greater afinity than it does with monovalent antibody.

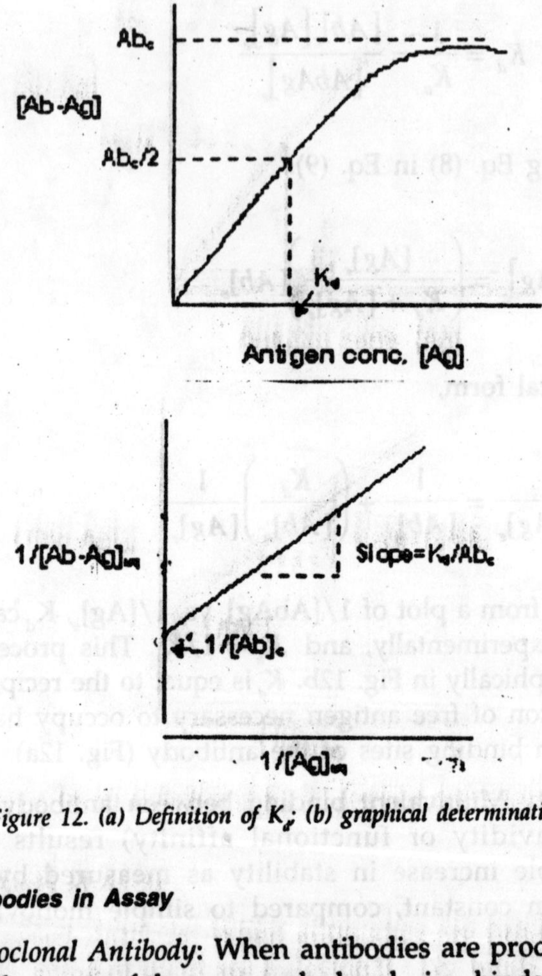

Figure 12. (a) Definition of K_d; (b) graphical determination of Ka.

Antibodies in Assay

Monoclonal Antibody: When antibodies are produced from B cells, some of them are polyclonal - i.e., binds with more than one type of antigen, whereas others are monoclonal - i.e., binds with only one type of antigen. Usually, antibodies are produced by immunizing an animal (such as a mouse). The antibodies produced are collected and the monoclonal antibodies are separated. The monoclonal antibodies (Mab) is the key for immunoassays.

ELISA: Although the antibody-antigen binding is highly specific, the direct detection of this measurement has been difficult. Therefore, labels are used for measuring the complexation product. Enzymes, fluorescent molecules, and radioactive labels are used for the labeling of the complexation. An example is ELISA (Enzyme-Linked Immuno Sorbent Assay). In this method, a known amount of enzyme-labeled antigen is added to a sample with unknown antigen concentration.

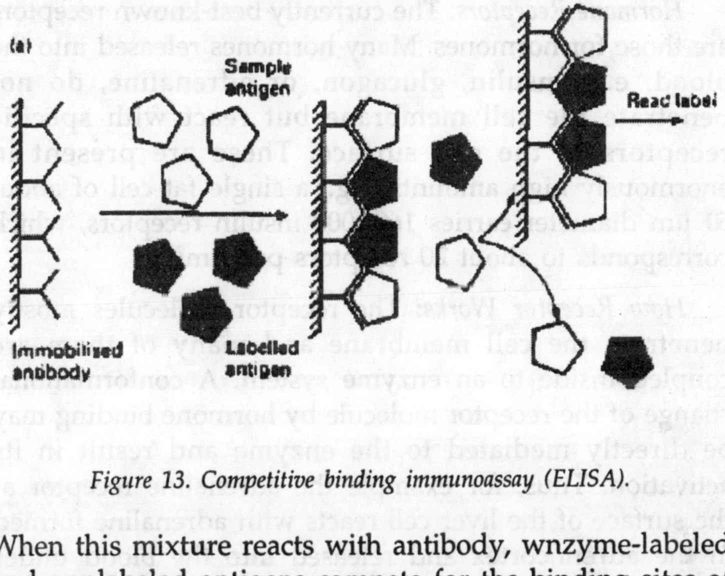

Figure 13. Competitive binding immunoassay (ELISA).

When this mixture reacts with antibody, wnzyme-labeled and nonlabeled antigens compete for the binding sites of the antibody. The more antigen there is in the sample, the lower is the fraction of enzyme-labeled antigen in the antigen-antibody complex.

After removal of unbound antigen the amount of bound enzyme-labeled antigen is determined via the enzyme-catalyzed reaction (Fig. 13). Widely used indicator enzymes are horseradish peroxidase and alkaline phosphatase. Urease and deaminases have also been used in immunosensors.

Receptors

Receptors at Membrane: Biological receptors are protein molecules with a specific affinity for hormones, antibodies, enzymes, and other biologically active com-pounds; most of them are bound to the cell membrane. A receptor-ligand interaction is transmitted to other molecules inside the cell, where consecutive reactions are triggered.

Hormone Receptors: The currently best-known receptors are those for hormones. Many hormones released into the blood, e.g. insulin, glucagon, or adrenaline, do not penetrate the cell membrane but react with specific receptors at the cell surface. These are present in enormously high amounts, e.g., a single fat cell of about 50 μm diameter carries 160, 000 insulin receptors, which corresponds to about 20 receptors per μm^2.

How Receptor Works: The receptor molecules mostly penetrate the cell membrane and many of them are coupled inside to an enzyme system. A conformational change of the receptor molecule by hormone binding may be directly mediated to the enzyme and result in its activation. Thus, for example the adrenaline receptor at the surface of the liver cell reacts with adrenaline formed in the adrenocortex and released into the blood under stress.

The resulting conformational change of the receptor molecule activates associated adenylate cyclase reaching into the internal space of the cell and converting ATP to cyclic adenosine monophosphate (cAMP). cAMP initiates the phosphate transfer from ATP to other enzymes by protein kinase. In this way a number of other enzyme reactions are started, leading to a cascade of activated enzymes. Finally, a single adrenaline molecule will have stimulated several thousands of enzyme molecules, which will themselves liberate about three million glucose molecules from glycogen within a few seconds. An extremely weak chemical signal is thus immediately enzymatically amplified a millionfold and the sugar reserve of the body is mobilized.

Smell Receptor: In addition to hormone receptors, taste and olfactory receptors are typical examples of this biospecific recognition process. Presumably there are about 20 to 30 primary smells. After being bound to the appropriate receptor, their molecules cause conformational changes in the receptor molecule leading to a depolarization of a part of the nerve cell membranes and initiating an action potential.

Light Receptor: Another receptor type is that of light receptors. The retina of the human eye contains about 108 tightly packed receptor cells. Here, biochemical reactions, namely of thc rhodopsin molecule, are not in-itiated by the binding of chemical substances but by light quanta. Thereactions are enzymatically amplified and transformed into electrical impulses via membrane potential changes.

Because of its light-absorbing chemical group the protein bacteriorhodopsin from salt-tolerant halobacteria has been studied in detail as a photoreceptor model. A single photon is sufficient to give rise to a conformational change of the protein and to transport two protons outside the cell. This 'proton pump' forms a proton and voltage gradient across the cell membrane driving the production of energy-rich ATP.

Current Status: Compared to the investigation of enzymes, that of the structure and function of membrane-bound receptors and their biotechnological application is only at the very beginning. An analogous classification, e.g., according to recognition mechanism or specificity has not yet been attempted. However, any progress in this field will provide impetus to the development of new biosensors based on receptors.

6

Visualisation Tool for Bio-Analysis

The role of bioanalysis applications is mainly to help scientists to discover the relevant hidden information inside databases. The bioanalysis area becomes a new domain of biological expertise and researches. One of the major stakes of life-sciences in general, and vegetable genetics in particular, as well as on the cognitive level applied in the fields agro-food and health, is the possibility of discover and describe genes of interest, and their regulations, determining the genetic factors from characters target expression for plants improvement in term of development, food value, resistance to the biotic and abiotic stresses.

Comparative Genomic and Bio-Analysis Concepts

A small number of cereals feeds most human and animal world populations, including wheat, rice, maize, sorghum and millet. Together with rice and maize, wheat is responsible for providing more than 60 % of calories and proteins needs. All these cereals belong to the Poaceae family, which includes more than 10,000 species, their common ancestor originated about 55 to 75 millions years ago. Among agricultural crops, the bread wheat (Triticum aesticum L.) has the largest genome size representing 16,000 Mb (40-fold larger than the rice genome), including chromosome 3B.

The wheat genome consists of seven groups of chromosomes (hexaploid form), where each group contains a set of three homologous chromosomes belonging to the A, B and D genomes. The A genome was contributed by Triticum urartu, a diploid wheat ancestor, and the B genome by an unknown close relative of Aegilops speltoides, another diploid wheat ancestor.

Finally, about 8,000 years ago, the D genome was added from a third ancestor called Aegilops tauschii to the AB genome tetraploid Triticum turgidum, giving the bread wheat has known today. Genetic mapping remains a necessary step to identify genes of interest. This is true for all crops, but especially in the wheat, since bread making quality is still today the most important trait to be improved.

Trait-directed sequencing is one of the ways available to identify these candidate genes for a giving Quantitative Trait Loci (QTL), located on a specific wheat chromosome locus for which compilation of all available data in terms of cDNA, EST (Expressed Sequence Tag), BAC, microsatellites and RFLP (Restriction Fragment length Polymorphism) markers is necessary.

Moreover, one of the great value of wheat genetics is the creation of a series of lines, known as nullitetrasomic and ditelosomic, which have been extremely useful in assigning markers to chromosomes and chromosome arms, respectively. Today, a higher level of resolution can be obtained by using a series of deletion lines. These lines having deletions of chromosome segments (deletion bins) can be used to assign also markers, but to particular regions of chromosomes, allowing links to be made between genetic and physical maps.

First comparative genetic mapping in grasses have shown, at the macro-colinearity level, that the gene order is well conserved along the grass chromosomes, despite large differences in genome size, chromosome number, ploidy level, and content of repetitive elements. However, recent comparative studies of large stretches of BAC DNA sequences at orthologous loci, in barley, sorghum, bread wheat, maize and rice have shown numerous colinearity exceptions at the micro-colinearity level.

These studies have shown that rearrangements caused by insertion/deletion of transposable elements, duplication, insertion, deletion, and inversion of genes, as well as gene movements have shaped the different grass genomes since their divergence from a common ancestor. Today, the use of DNA sequence-based comparative genomics for evolutionary studies and for transferring information from model species to related large-genome species has revolutionized molecular genetics and breeding strategies for improving those crops.

Comparative sequence analysis methods can be used to cross-reference genes between species map, enhance the resolution of comparative maps, study patterns of gene evolution, identify conserved regions of the genomes (colinearity), and help interspecies gene cloning. Rice (Oryza sativa L.) is commonly chosen as an excellent model within the grasses to study comparative genomic, because it has a relatively small genome among the Poaceae family and mostly genome compared with bread wheat.

The availability of DNA sequence information for the rice genome, especially the availability of the 12 pseudo molecules, and deletion map position of a large number of wheat loci detected by EST unigenes allow to obtain a detailed picture of "orthologous" relationships between wheat and rice chromosomes.

Therefore, biological data, and DNA sequence data in particular, are accumulating at a phenomenal rate. It is obvious that the achievement of complete plant genome sequencing (Arabidopsis, rice) is only the beginning of structural, functional and evolutionary studies of plant genomes. Visualization methodologies are being applied to discover and represent this knowledge in a form that human easily understands.

Virtual Reality Concepts (VR)

There are several synonyms that define Virtual Reality in the literature: Synthetic Environments, Cyberspace, Artificial Reality, Simulation Technology, etc. Despite the recent developments, "Virtual Reality is not a new concept". The oxymoron "artificial reality" was introduced by Krueger in 1983 and Sutherland published in 1965 an IFIP congress proceedings with all the key concepts of immersion in a simulated world, as well as a complete sensory input and output, which was and still is the basis of VR research.

A virtual or simulated world is created and based on three-dimensional graphs and audio elements that can be experimented in real time. Virtual Reality techniques try to minimize the barrier between simulation and the user. In VR applications there is the notion of "freedom of ride", where the user can choose the angle or better position to observe the system, not restricting himself to just some points of pre-defined views.

Beyond this point, placing a user in a simulated environment is usually far more economic than putting him in the real physical environment. Virtual reality is the generic name under which are being grouped all the means by which a user can, in real time, freely explore, examine, manipulate and interact with extremely complex information in a computerized world.

Virtual Reality: Immersive and Non-Immersive

It is very important to distinguish between VR immersive applications and VR non-immersive applications. Behind this concept, there is a big discussion about what it is or what is not VR. Despite all discussions, particularly, non-immersive VR is defined as the visualization obtained trough a window, typically the computer monito·.

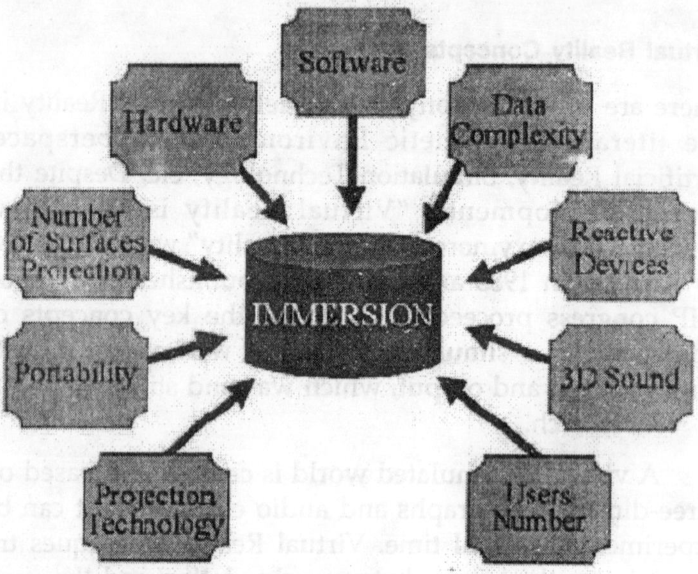

Figure 1. Main Factors contributing to immersion sensation

There are many reasons to invest in this technology since many home computers around the world are potential targets. Furthermore, the end user only needs to install a free scene viewer, web browser or viewer plug-in software to visualize almost all VR applications available on the Internet. By Immersive application we mean an application that allows the coupling of almost all user senses, mainly the vision, audition and tactile senses.

Full immersive environments are known as Caves, and other systems have less degrees of immersion. Many factors can contribute to immersion sensation as showed in figure 1. Santos proposed a classification of immersive visualization tools: desktop system, desks systems, wall systems, room systems, and caves systems.

Non-Immersive VR Technique

The X3D/VRML is an extensible open file format standard for 3D visual effects, behavioural modelling and interaction that allows the creation of 3D scenarios, where a user can stroll, visualize objects under several angles, and yet interact with them. The VRML language was designed to describe interactive simulations of composite participants, in virtual worlds available on the Internet and tied with World Wide Web.

A VR application designed for the Internet provides an intuitive and common interface, which facilitates interactions with 3D objects; this is one of the reasons for the recent fast acceptance and utilization of VR. Dealing with the Internet history, the Web standard and its evolution was quickly accepted since companies and universities world-wide are continually developing it.

The Web is still currently the subject of much interest to both simulation researchers and simulation practitioner's. To navigate in virtual worlds using the X3D/VRML standard, a web browser or a viewer plug-in. Commercial, freeware and GPL-style licenses are available for all the latter mainly because a considerable number of companies and universities are involved in the X3D/VRML development and thus they created their own viewers.

This diversity helps in providing creative solutions to improve the most crucial matter - the user interface; thus the user can more easily select the viewer fitting his needs. According to Fishwick, "...The VRML plug-ins allows one to construct 3D geometry and have it rendered in real time. A document contains an URL (Uniform Resource Locator), which points to a VRML file. The browser recognizes this particular information type and launches the VRML plug in. Then the user browses a 3D scene using whatever HCI (Human Computer Interface) devices are available...".

The X3D/VRML has a set of characteristic that turns your computer in a powerful and efficient visualization tool (among them, the most important that stands out are: cost, portability, hyperlink, usability, spread, cooperation). Common application fields are in the following domains: medicine, computer aided design, modelling & simulation, scientific visualization, Geo-Spatial, education, entertainment and technical training. In addition, users will not need any special training or previous experience to use a VRML viewer, the navigation controls helps walking through the model, flying over it.

"...The quality of an interaction depends greatly on its responsiveness...". One of the keys to the success of visualization tools is the ease of use. Scientists whose interest is to find meaningful in data - have neither time nor patience for graphics subroutine packages, computer graphics jargon and all sort of other unnecessary details. In August 2001, the Web3D Consortium, the task force group that develops the X3D/VRML standard, launched the X3D open standard as a new-generation successor to VRML to bring rich and compelling 3D graphics to the Internet for a wide variety of applications and devices.

The X3D is being developed under the Web3D Consortium's standardization process that provides a full and open access to the specification to interested companies and eventual submission to the International Standards Organization (ISO) for ratification to provide long-term stability for Web3D content and applications.

Development, Stages, and Strategies

The first objective of this application is to allow better biological interpretations from the study of the 3B chromosome. The purpose is specially to seek the macro and micro-colinearity zones between rice and wheat genomes, in order to help the stretches of BAC DNA sequences for the development of a physical map for this chromosome. The application focus is to allow putting this entire data collection set with its features and correlations inside a VR environment, where no user skills in computer graphics are needed to manipulate them.

Today, the Internet environment is usual and quotidian to almost all biological scientists mainly due to:

a) the access centralization to updated genome databases, such as NCBI, EMBL, SWISSPROT;

b) the great number of tools and servers dedicated to sequence search, 3D homology modelling, and so on;

c) a collaborative environment among several international research groups physically distant;

d) the velocity of communication, data and information updating.

The X3D/VRML technology was chosen as a development platform for this application since:

a) the user interface (navigation, examination, etc.) is built inside available viewers;

b) there are different viewer licenses available: commercial, freeware, and GPL-style;

c) the application development is fastened since it focuses on the scientific data to visualize and not on programming;

d) it represents an evolutive Internet ISO standard (with many international research groups working, collaborating, and improving this standard, including many universities and enterprises);

e) the international support is available through many mailing lists and FAQs (Frequent Ask Questions); and

f) this standard was originally created to work fine with the new biologist environment (Internet).

This interdisciplinary collaboration project was structured and composed in several stages: the establishment of methodology work, the geometrical modelling, and the user interface modelling.

Interdisciplinary Collaboration

First, biologists and computer science professionals have different methods of thinking, different languages, and different methodological ways to work. Therefore, the first basic stage was the establishment of a methodology of work. The language standardization is not an easy task to realize. So, the strategy adopted was the creation of a memory document that has been updated after each meeting, including all decisions taken. This strategy has minimized several communication misunderstandings.

Geometrical Modelling

The second stage focused on the problem knowledge and on how to organize all information inside the input data. Two different files were used to build this application: the parameter file and the data file. The parameter file allows the establishment of the main object structures that will represent the wheat chromosome analyzed, the deletion bin (inside the wheat chromosome structure viewed by cytogenetics), and the twelve rice pseudo-molecules (virtual chromosomes).

The data file contains all macro and micro-colinearity relationships based on similarity between genetic markers sequences (named "query" hereafter) and the rice full sequence pseudo-molecules (size alignment, pseudo-molecule coordinates, Evalue value). A query is located within a wheat bin term named hereafter "deletion bin". The similarity relationship is obtained using the BLAST (Basic Local Alignment Search Tool) algorithm.

Between all algorithms available, we have chosen a BLASTn (nucleic against nucleic) algorithm using the default parameters, since we need to find true biological relationships (sequence alignments) at the DNA level. The origin of the wheat markers sequences used for the BLASTn query analysis are molecular markers (cDNA, ESTs or genomics) assigned to a given deletion bin for a given wheat chromosome using a series of wheat deletion lines.

Wheat deletion lines are specific wheat varieties, which have a peace of a break out chromosome arm. Every query sequences (wheat genetic markers located within a deletion bin) may give a hit against the rice pseudomolecules, that means that this specific wheat marker recognizes a specific rice pseudo-molecule. Then, for each hit, a query sequence may reveal many HSP (High-scoring Segment Pair) along the rice pseudo-molecule corresponding probably to several rice putative "genes" or relevant sequence similarity.

Each HSP represent a specific DNA sequence alignment corresponding to a given Evalue or Expectation value that is the number of different alignments observed with a score equivalent or better than a given threshold expected to occur in a database search by chance. The Evalue and its significance score work inversely. So, lower is the first, bigger is the second, and consequently greater is the HSP relevance. The main BASTn parameter for which we can define a minimum cut off is the Evalue. Therefore, here used an Evalue cut off of 10^{-1} that means that all the value above 0.1 are discarded. The values below the cut off are classified in 5 categories:

— Class A: Evalue bellow 10^{-1};

— Class B: Evalue bellow 10^{-5};

— Class C: Evalue bellow 10^{-25}

— Class D: Evalue bellow 10^{-50}; and

— Class E: Evalue bellow 10^{-100}.

The geometrical modelling stage was concluded after parsing the two input files generating all 3D elements. A scale adjusting step was needed to harmonically visualize all information, because the biological units are measured in base pairs (bp) for physical maps, percentage for cytogenetic maps, and centimorgan (cM) for genetic maps, but their proportion and graphic representation in this application are measured in meters.

User Interface

This stage represents the most complex task developed in this application because all data correlations and its behaviours possibilities were implemented. The main features provided by the user interface are:

— *Names of deletion bin*: It is possible to visualize the names of each deletion bin when the user passes the mouse cursor over it inside the wheat chromosome graphic representation. This option enables the biologist to know precisely what wheat markers will be activated or deactivated. To implement this application feature, we have associated each deletion bin graphic representation to its respective name inside a hyperlink description field with a null URL.

— *Selection of a deletion bin*: After selection there is a colour change inside the wheat chromosome. The corresponding HSP on rice (one or several) will be turned on with the same corresponding deletion bin colour. This option enables the biologist to view similarities between the sequence of a specific wheat marker and a rice locus on the pseudo-molecules, according to the Evalue.

The implementation of this feature is done this way: each selection (with a mouse click) of the deletion bin graphic representation implied a change in the colour and transparency properties of its respectively HSP. However, the sequence alignment inversions between the wheat and rice are not displayed at the graphical level.

- Selection of all deletion bins at the same time, just by turning on the "Light All (on/off)" option. This option simplifies the complete visualization of all deletion bins as well as the corresponding HSP that will be visible with the same colour as the deletion bins (figure 2). If a deletion bin is already turned on when this option is turned on, all deletion bins minus the last one, will be turned on, according to the Evalue.
- *HSP Size*: After selecting a deletion bin in the wheat 3B Chromosome, we can access to precise visual information if the user changes the position of "Scale Hit Rice" slider. At start time the HSP representation are very small in size inside the rice structure, and then it is not always visible. Therefore, the user can increase its Y size using a slider to see precisely in which rice chromosome we have the corresponding HSP.
- *Evalue Threshold*: After selecting a deletion bin, we can use an "Evalue" slider. Each HSP on rice has its particular Evalue and at start position all HSP are potentially visible. If the user changes the position of this slider, the Evalue belongs to an interval (5 classes intervals are retained: A, B, C, D and E), and all HSPs

that have an Evalue in the same class interval are displayed.

This option enables the biologist to change very easily the threshold of each interval and then he has the opportunity to identify more or less relevant HSPs. The transparency property value of selected HSP graphic representations is used to trigger the display according the position of the "Evalue" slider.

— *HSP query name*: After selecting a deletion bin, we can visualize the HSP corresponding query name when the user passes the mouse cursor over it inside the rice structure. This option enables the biologist to know with more precision the interesting target to analyse. Like for the display of deletion bin names, we implemented this application feature, by associating each HSP graphic representation to its respective name inside a hyperlink description field with a null URL.

— *Wheat marker display*: After selecting a deletion bin, select a HSP and then turn on the respective wheat bin markers inside the corresponding deletion bin. This option enables the biologist to evaluate the starting from a zone of interest in the rice in correspondence to the specific wheat zone concerned. This application feature is achieved by changing the colour and the transparency properties of wheat bin markers on each mouse click over the HSP graphic representation.

— *Scaling on wheat marker*: Changing the position of the "Scale Query wheat" slider enables to increase the wheat marker size. At start time the wheat bin markers graphic representation are a very small inside the deletion bin graphic representation. Therefore, the user can increase the size, the Y size property of wheat bin markers is associated to the position of this slider.

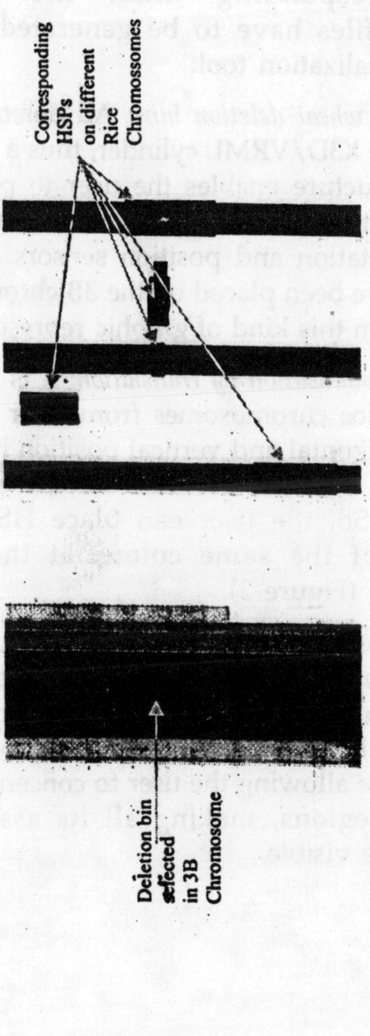

Figure 2. Selection of Deletion bin with all respective HSPs placed side-by-side after their displacement, at maximum scale and using the zoom feature.

— *Wheat marker Inspector*: After a HSP selection, it is possible to obtain more information about the respective wheat bin marker that has been turned on if the user touch it with mouse click. This click launches a HTML window opening inside the Web browser. For each wheat bin marker a hyperlink is set to the corresponding ".html" file. All the corresponding files have to be generated before running the visualization tool.

— *Easy access to all wheat deletion bins*: All deletion bins are placed over a X3D/VRML cylinder, thus a rotation of the wheat structure enables the user to place the deletion bin of interest in front of his rice chromosome selection. VR Rotation and position sensors (vertical & horizontal) have been placed on the 3B chromosome structure to obtain this kind of graphic representation.

— *Correspondence visualisation by translation*: It is possible to translate all rice chromosomes from their original position to a horizontal and vertical position just near the region of interest in the wheat graphic representation. So, the user can place HSPs and deletion bins of the same colour at the same observation level (Figure 2).

— *Zoom*: For all the features previously cited, the VR viewer enables any kind of zooms within the graphical structure domain. This option is not specific to our implementation, but it constitutes an additional and interesting feature allowing the user to concentrate on small specific regions, making all its associated information more visible.

7

Emerging Biotechnology Industries

The ability of developing countries to benefit from advances in biotechnology remains controversial. Some believe that biotechnology will not benefit developing countries in any way, while others argue that biotechnology will contribute significantly to the development of poor nations. There is evidence that biotechnology capabilities are concentrated in a few countries and that biotechnology remains a research-intensive field.

Therefore, developing countries that lack intensive research capabilities are unlikely to be part of this technological revolution. Furthermore, most of the products currently on the market and those in the pipeline are targeting the markets in developed countries and those with economies in transition. It is also evident that a few developing countries are becoming active participants in the new bioeconomy.

These developing countries present beacons of hope from which other countries could tailor their programmes to become participants in the new bioeconomy. At the same time, products that target markets in developed countries may be useful for development just as the mobile phone technology has been widely accepted in poor countries. The extent to which developing countries become participants may depend on their ability to adapt, innovate and use biotechnology products and services safely. It may also be influenced by international regulatory regimes governing biotechnology and the usefulness and applicability of the innovations.

Developing countries face a daunting, but not impossible, task in becoming real players in the new technology-driven economy. The greatest challenge lies, firstly, in amassing sufficient human capital capable of sustaining scientific enterprise, secondly, the political will and management foresight required to harness the developing technologies and, thirdly, having the ability to seek unique but efficient research, development, production and marketing strategies.

Similarly, the legal and regulatory regimes that would promote technology development and safeguard the public interest have to be set up and cultivated. These technological advances will bring new opportunities, benefits and risks, just like the current technologies. They may help increase the life span and the quality of life.

For example, ailing organs that shorten life could be replaced and diagnostic systems could be improved, be accessible and available. It will also be possible to stop some infection by blocking mechanisms by which disease-causing agents enter the body, proliferate and colonise body tissues. These opportunities are unlikely to be evenly distributed but will have a positive impact on all.

For example, vaccines could be administered as a single treatment through the use of drug delivery devices, cutting down the need for two to four visits to clinics to get boosters. Combined with increased information availability, reduction in cost, complexity and energy needs of diagnostic systems, even village clinics may benefit.

However, it is also important to address the policy issues related to these technologies as they develop. For example, human cloning is one subject where the rights, ownership and integration of clones in human society remain controversial and unresolved. Similarly, the rights, choice, incentives and selection of surrogate mothers that would carry these clones may be controversial too.

As tools get smaller, smarter, more productive and efficient, the mass education of entire populations will assume new proportions. Knowledge will be easy to access, systems easy to configure and produce. Consequently, the responsibility of individuals will increase. Similarly, the chance of abuse of any of these technology platforms will increase proportionately. By the same token the security of individuals will increase as detection tools and protective materials become cheaper. Managing the benefits and risk of technology will be part of the future governance challenges.

Biotechnological Innovation

The last century saw the replacement of plant-derived products with petroleum derivatives for energy and chemical product generation. These remarkable transformations helped humanity to overcome some of the natural limitations of relying on natural processes. The change was largely a result of advances in chemistry and allied fields. This century promises to open new avenues for increasing the use of renewable resources in the global economy.

These trends will open up new opportunities for the participation of developing countries in the new bioeconomy. But as in previous technological revolutions, the promise and reality are different. In the case of agricultural biotechnology, for example, only a handful of developing countries have so far managed to become players in the global economy. The rest have little hope of playing significant roles in the near future. As in other technological fields, participation in the new bioeconomy will be uneven and limited to those countries that make the necessary investments in technological development. So far, much of the research on policy aspects of biotechnology has focused on agricultural and pharmaceutical biotechnology.

The field of industrial and environmental biotechnology remains understudied. Industrial biotechnology covers two distinct areas. The first area is the use of renewable raw materials (biomass) to replace raw material derived from fossil fuels. The second is the use of biological systems such as cells or enzymes (used as reagents or catalysts) to replace conventional, non-biological methods.

Industrial application of biotechnology is emerging as a spin-off from developments in other fields such as the pharmaceutical sector. This emergence is largely because industrial biotechnology has not received the same level of public policy attention, as has biotechnology in other sectors. There are other structural factors that influence the diffusion of industrial biotechnology.

These include the dominance of physical and chemical technology as a source of concepts for the design of industrial plants that are important in introducing biological processes. One of the main advantages of industrial biotechnology is the prospect for the controlled production of biological catalysts. These biocatalysts are more specific and selective than their non-biological counterparts. As a result, they offer greater potential for cleaner industrial production.

In other words, biocatalysts generate fewer by-products and can start with relatively less purified feedstock. And because they are self-propagating, they can be used in applications such as waste treatment (OECD, 1998) But despite these advantages, biocatalysts are generally fragile (requiring large amounts of water) and have low volumetric productivity. Over the years, however, incremental technological innovations and new bioreactor designs have helped to improve the industrial performance of biocatalysts.

With incremental improvements in biocatalysts and the emergence of new design concepts, biotechnology's capacity to diffuse in the industrial sector will be enhanced. This prospect is enhanced by the growth in the biological sciences, as well as complementary fields such as chemistry and informatics. The use of biomass for energy and industrial uses has been on the agenda of many Governments for nearly two decades. Much of the interest was triggered by the oil crises of the 1970s.

Although interest in the field waned with the decline in energy prices, advances in the biological sciences have continued to enhance the prospects for technological improvement and wider application. In addition to energy, living plants can be used to produce chemicals such as citric acid, lysine and lactic acid. Genetic modification offers new possibilities for using plants as a source of raw materials for chemicals or even finished products.

Monsanto, for example, has experimented with a genetically modified crest plant to produce a biodegradable plastic using a gene from a bacterium, Ralstonia eutropha. Similar experiments are underway in other chemical firms around the world. One of the most advanced efforts is an initiative by Cargill Dow Polymers (CCDP) to construct a plant to produce 140,000 tons a year of polylactide (a biodegradable plastic) using lactic acid fermented from corn.

As enzyme technology improves, attention is shifting to other areas of bioprocessing by tapping the potential in the world's wealth of microbial life. With the advent of DNA sequencing, microorganisms will become an important addition to industrial activities through scientists' discovery of new biocatalysts. The field of genomics is therefore likely to extend its influence from medicine and agriculture to industrial production

Methods such as forced evolution and rational design will increasingly be used to discover new enzymes for industrial use. In addition, methods suc h as gene shuffling are helping firms to optimise their bioprocessing activities. It is expected that the genomes of major industrial microorganisms will be sequenced in the coming years, and this will add significantly to the library of industrial biotechnology. Prospecting for biological organisms of industrial value will increase as bioprocessing gains acceptance.

The network of agreements between bioprospecting firms, such as Diversa, and biotechnology-related firms such as Dow, Aventis, Glaxo, and Syngenta, illustrates the growing interest in this field. These technological developments will result in new generations of chemicals and polymers that will compete directly with bulk petrochemical products. These developments are improvements over a long history of efforts to use bioprocessing in industry. The mining sector, for example, has been using bioprocessing for a long time.

Bioleaching is a natural process that helps in weathering of sulphide ores. Organisms have been harnessed successfully in zinc, copper, nickel and gold mining among others. Bioleaching is environmentally friendly, less costly to build and cheaper to run than traditional leaching methods.The use of Thiobacillus ferrooxidans to oxidize sulphide metal ores dates back several centuries. But it was not until 1986 that the world's first commercial bioleaching tank for gold-bearing sulphide was commissioned in South Africa by Gencor. The procedure now employs both mesophilic and thermophilic microorganisms.

Developing countries have been the centre of origin for some of the bioleaching technologies. For example, the Biox process owned and developed in South Africa is now used in many other countries, while the success of the largest experimental solvent extraction electro-wining (SXEW) in Zambia has been used in 40 other mines in developed countries. Current efforts are focusing on identifying new organisms that are stable, multiply efficiently and are reliable.

Other initiatives focus on identifying organisms for different processes suc h as reduction of iron and sulphates, as well as those that can be used to dissolve toxic metals in mine waste discharge. Technology is also being developed to inhibit organisms that are involved in mine acid generation. Biofilters for sulphides represent an effective and inexpensive alternative to traditional filters.

The textile and leather industry is another sector that has used biotechnology for decades, especially through the use of enzymes. Firms such as Maps (India), Novozyme (Denmark) and Genencor (United States) are marketing many enzymatic products (amylases, lipases, cellulases, isomerases and extremophiles). Genetic probes have been developed that could detect adulteration of the merchandise and deterioration of fibre.

There is now a movement towards preparation of high-quality fibre from microorganisms, plants and animals, using genetic engineering technology. The demand for high-quality leather and fibres is likely to increase. Developing countries could add value to their raw materials and meet demands by adopting the new technologies in their processes. The leather and textile industry is faltering or dying in some countries, owing to the poor quality of products and increased costs of production. Biotechnology, if used appropriately, could lower the cost of production and improve the quality of leather and textile products.

The paper industry is another old technology that relies heavily on wood, energy, water and chemicals. New technologies have emerged that are changing the face of this industry. The pulp and paper industry was estimated in 2000 to be the fastest growing market for industrial enzymes. Enzymes are quickly replacing traditional chemicals in pulping, in paper production and in de-inking recycled paper.

Biopulping (using fungi) results in a nearly 30 per cent saving of electricity, while treatment with cellulase and hemicellulase reduces wood-drying time considerably. Bioleaching of pulp reduces chemical requirements by 50 per cent. The use of enzymes and fungi increases the physical properties of the fibres and the quality of paper. Many developing countries have lagged behind in technological developments in paper manufacturing and have become importers of paper, even when they have the potential to be exporters.

In addition to industrial applications, modern biotechnology also is likely to make major inroads into the field of environmental management. Using microorganisms or their products, environmental biotechnology involves processes that detoxify industrial waste, clean up industrial contaminants and enhance control of environmentally unfriendly practices.

Environmental management is indeed a growing industry that will benefit from further advances in biotechnology. The use of sensors to detect acidity, electricity, ionic strength, heat, light and smoke, among others, has been the basis of many analytical and home instruments. The selectivity of enzymes, other proteins, nucleic acids and carbohydrates has become the focus of research to identify noxious gases in mines and organisms in the environment.

Characterisation of Biotechnology Industry

There is no single characterisation of the biotechnology industry in developing countries. What is evident, though, is the growing importance of international partnerships and alliances in biotechnology's evolution in developing countries. Such alliances are likely to increase with time. These alliances also serve a larger function: they provide a basis for the kinds of partnerships that are essential for a market inclusion model to function.

The absence of open markets for technology makes technology acquisition difficult. For example, vaccine and drug discovery technologies may not be easy to acquire, as there is not one such technology on the market. However, those that have built significant technological capability (human and facilities) may be seen as denying those that do not have similar capacity the benefits of new advances and could easily become a source of resistance.

Thus, formation of partnerships between those that have the capacity and those that do not may be essential for the effective commercialisation of biotechnology products. A winner-take-all scenario that was tried in the field of agricultural biotechnology is slowly giving way to opportunities that explore new models of international cooperation. So far, much of the public debate on this issue has tended to focus on the importance of partnerships between the private and public sectors.

While these linkages are important, they are not a substitute for international alliances, under which the private sector plays a critical role in the commercialisation of new products.

Commercialisation of Biotechnology

Biotechnology in developing countries has been promoted through government programmes. However, biotechnology research and development are mainly in the hands of private firms in developed countries. Unless developing countries are able to induce private sector participation in biotechnology, alliances between private firms from developed countries and public or private institutions in the developing countries will be viewed with suspicion.

Developing countries are slowly but steadily adopting transgenic products. The number of countries growing transgenic crops has grown from 3 in 1996 to 8 in 2001. Similarly, the area planted with transgenic crops in developing countries has grown from 1.3 million hectares to 14 million hectares over the last 6 years. Argentina accounts for about 80 per cent of the acreage in developing countries and the People's Republic of China has recorded significant threefold growth in acreage in 2001.

The growth in area (in millions of hectares) planted with transgenic crops in developing countries between 1996 and 2001 is shown below. Plants are at the centre stage of the modern biotechnology revolution for production of pharmaceuticals, fine chemicals, industrial enzymes and other products that are not necessarily food or feed.

Unfortunately, the role of plant biotechnology as a source of renewable fuels, degradable plastics, rubbers, adhesives and other products derived from fossil fuels, in addition to industrial and food products, is buried in the debate over the safety of genetically modified organisms for human health and the environment. Some drugs, such as aspirin and menthol, that are currently in use were derived from plants but are synthesised chemically to meet economic and quality considerations.

Biotechnology may offer alternative production systems of plant products by either boosting levels of the desired ingredient(s) in the plant or improving the efficient recovery and quality of final product. This field is still developing and poor countries with excellent growing conditions for tobacco, potatoes and corn, among others may become the future home for biofarming centres.

Countries with the capacity to purify, produce and package these products may have an added advantage. The national regulatory regimes and international governance policies of transgenic crops engineered to produce drugs are still developing. The ability of these regulations to enable participation of poor nations will depend upon their flexibility, responsiveness and inclusiveness.

Development of Biocon

In India, the importance of international alliances is illustrated by the evolution of Biocon India. The company was established in 1978 in Bangalore as a joint venture between Biocon Biochemicals of Ireland and local interests. The company started with the production of simple ferme ntation products and later embarked on its own R&D programme that saw it become a major player in the fields of modern biotechnology.

The R&D efforts were inspired by the need to diversify the company's product portfolio. One of the first efforts was to develop a local alternative to Konji-a carefully fermented mass of cooked soybean meal and roasted wheat-imported from Japan. This is a good source of amylases and proteases, enzymes crucial in the hydrolysis of carbohydrates and enzymes.

The process was complicated and largely unknown outside Japan. But after three years, Biocon India successfully mastered the techniques, leading to new fermentation platforms and enzymes that matched those from Japan. These successes were also encouraged by Biocon Ireland, which bought products from the company. Biocon India became an owner of new fermentation technologies, and two manufacturing plants were commissioned to meet demand.

Following these successes, Biocon India became a supplier of food enzymes to United States and European markets. In addition to enzyme production, the company also invested in the development of new production systems that incorporated the advantages of solid state and submerged fermentation. After five years of effort, the team developed Plafractor, a solid phase fermentation platform with automated and programmable controls, allowing reliability, repeatability and reproducibility.

It is also a closed system that protects the operator and the environment from any toxic agents produced during fermentation. Further, it allows the quick and convenient recovery of fermentation products and saves space and labour. The patented bioreactor recently won the 2001 Biotechnology Product and Process Development Award from the Indian Department of Biotechnolo gy in the Ministry of Science and Technology.

The success story of Biocon India is an example of the importance of international partnerships. While Biocon India carried out innovations and production, Biocon Ireland provided the market for its products, enabling the newly formed firm to have a steady flow of income as well as eliminating marketing costs of products. In 1989, Unilever acquired Biocon Ireland and its 30 per cent share in Biocon India. Unilever's financial muscle and global standing gave Biocon new linkages and access to funds.

Biocon learned Unilever's global operations, standards and financial methods. Biocon India has expanded its operations and ventured into fields such as pharmaceutical research. It established Syngene, which in turn spun off Clinigene International as a wholly owned subsidiary. Syngene has close collaborations with AsraZeneca, Glaxo and BMS, which contribute to its research efforts. Clinigene specialises in genomics and clinical studies to support the pharmaceutical section of the Biocon Group. Biocon has thus developed rapidly through strategic partnerships with end users of its products.

Research, to a great extent, has been driven by the demand of its customers, resulting in accumulation of proprietary technology and development of products and processes. This flow of information between producer and end user was an important input in R&D activities. Biocon India went through a steep learning curve in global management, standards and negotiations. It exploited the chances that were presented to it, through association with global companies, to expand its markets and product range.

All these lessons helped the company to consolidate its position, identify funding opportunities and take advantage of market availability. The creation of different units into individual companies spurred their expansion, depth of research and product development. The autonomy enjoyed by the different units soon led to innovations that became the basis of new companies and new associations with other companies outside the group.

Biocon India represents one model of biotechnology commercialisation that depends largely on international partnerships and alliances. It carries with it the attributes of inclusion that should be encouraged in the development of industrial and environmental biotechnology.

Development of Herber Biotech

In Cuba, the case of market inclusion through international alliances is also illustrated by Cuban experiences in biotechnology commercialisation. The Center for Gene tic Engineering and Biotechnology (CIGB) has about 192 laboratories in total, equipped with the best instruments from countries such as Japan, Germany and Sweden. These facilities produce vaccines for meningitis B and hepatitis B. Vaccines for HIV, haemophilia and cholera are under development.

In diagnostics, CIGB has produced analytical systems capable of detecting HIV, hepatitis B, herpes simplex, Chagas, leprosy and other diseases. It has also produced probes for plant diseases, about 50 enzymes (some of which are produced only in Cuba), and 160 medical and pharmaceutical products.

Cuba moved into the commercialisation of biotechnology products through the creation of a semi-private enterprise, Herber Biotech. By 1998, Herber Biotech was recording about $290 million in sales of hepatitis B vaccines and its pharmaceuticals in 34 countries. The company also had representatives in about 50 countries. Nationally, biotechnology was placed just behind tourism, nickel production and tobacco in terms of export earnings.

The company is extending its partnerships with other developing countries. For example, in 2001, it established a joint venture with Kee Pharmaceuticals of India, in Haryana. This marketing venture is aimed at getting access to the Indian market through special pricing and technology transfer. Some of the company's main products, such as a recombinant of streptokinase called cardiostrep, is used for the hydrolysis of coronary clots or prevention of heart attacks and has a potential market value of about $11 million per year.

The market value is expected to grow by 30 per cent annually. Other products involved included interferon and human transfer factor, also owned by Herber Biotech. Cuba recognises that participating in the global market involves forging alliances with a wide range of enterprises, especially those that have extensive marketing networks. For this reason, Pfizer, a US-based multinational firm, is marketing some of Herber Biotech's biotechnology products, such as the meningitis B vaccine.

Cuba's biotechnology industry is an example of the importance of political leadership on technological matters, domestic funding for research activities, creation of appropriate research institutions, and international alliances for product commercialisation. The future of the Cuban biotechnology programme will depend on the degree to which these elements are maintained, especially in the face of worsening economic conditions that might divert allocation of resources and political commitment to other sectors.

Cuba's success in biotechnology is attributed to the existence of a critical mass of scientists in natural and applied sciences as well as a political commitment to technological innovations. If the United States market opens up to Cuban products and foreign investment in Cuba picks up, then Cuba will reap more benefits than it currently does from its investments in biotechnology over the last two decades. If this happens, cooperation between Cuban scientists and their American counterparts is likely to increase.

Electric Genetics

In south Africa, the field of genomics also offers opportunities for developing countries to commercialise specialised bioinformatics services. The data derived from genome sequencing activities need to be analysed to determine the functions, activity and regulation of genes and facilitate product development. Traditional methods of test tube analysis of each molecule's chemical, physical and biological properties are laborious and insufficient when faced with thousands of different molecules from various organisms. As a result, firms devoted to bioinformatics are starting to emerge in developing countries.

For example, in 1996 Electric Genetics was established in South Africa to commercialise innovations from the South African National Bioinformatics Institute (SANBI). Electric Genetics is housed at SANBI at the Western Cape University and is funded through a national innovation fund aimed at encouraging science, engineering and technology.

The company and its partners, SANBI and Silicon Graphics, have developed two software packages for clustering, alignment and recombination of sequence data. The first programme is a management system for expression variation analysis and transcript reconstruction, while the second is a sequence tag alignment and consensus knowledge database.

Leading genomics institutions such as Celera, the Institute for Genomics Research (TIGR), Paracel, Paradigm Genetics and the German Cancer Research Center have used the programmes. In 2000, Electric Genetics won the prestigious Technology Top 100 Award in its category from the South African Ministry of Trade and Industry. The award was a recognition of the company's excellence in development and research, as well as of its ability to commercialise its products internationally.

International partnerships have played a key role in the development of the company. Electric Genetics may be said to be the marketing arm of SANBI. SANBI is part of the Western Cape University. This association with a local university has given it a unique position. SANBI offers post-graduate courses in bioinformatics and has conducted training for medical representatives in Southern and Eastern Africa in conjunction with Tropical Diseases

Research (TDR) arm of the World Health Organisation (WHO). Electric Genetics may be said to be a spin-off of a university.

Nucleotide Sequencing and Analysis

In Brazil, genome sequencing is a highly specialised biotechnology field that has been largely covered by large organisations such as the Institute of Genomic Research (TIGR) and the Sanger Center. Genome sequencing requires a high level of sophistication and organisation in the preparation and tracking of samples and analysis of the voluminous genetic data developed.

However, the Organisation for Nucleotide Sequencing and Analysis (ONSA) captured the headlines when it announced that it had successfully completed the sequencing of Xylella fastidiosa, an organism with a genome size of 2.7 megabases that infects oranges and causes Citrus Variegated Chlorosis, 2 months ahead of time and $2 million within budget. They had, within the same period, built a genomics institution from scratch that was unique: a virtual institute involving some 34 laboratories located in geographically distant places and belonging to different institutions.

When the São Paulo State Research Support Foundation (FAPESP) decided to fund the genome sequencing project, it also wished to expose as many laboratories as possible to the modern biotechnology tools. Plans to create a single centre were rejected. They settled for a virtual institute that was web (Internet) managed. It was also cheaper than building one large concrete block institute. For these reasons, the project is estimated to have trained at least 200 young geneticists.

FAPESP is entitled, by law, to 1 per cent of tax collected by the State of São Paulo. Similarly, FAPESP is not allowed to spend more than 5 per cent of its budget on administrative duties. Given the success and independence it has enjoyed over the years, the genome project has provided another opportunity to score a word's first in a unique fashion. Indeed, Brazil became the first country to sequence completely a plant pathogen. From its budget of $250 million in 1998, FAPESP is estimated to have spent about $45 million since 1997 on the ONSA project.

The initial $11.6 million budget helped established two central sequencing laboratories and a bioinformatics unit, while all the other selected laboratories received the necessary equipment. Rather than appoint, the project requested interested laboratories to send in bids upon which selection was based. The project developed rapidly. The idea of bacterial genome sequencing was put forward in May 1997. By the mid-November, 1997 the project had been announced, the project structure defined, the steering committee formed, the organism chosen and the participating laboratories selected.

The first sequence was deposited in March 1998 and the sequence completed in February 2000. The initial success of the project attracted some of the nationals based abroad to return to Brazil. The choice of the organism was based on its economic importance, industrial interest and size of the genome. The organism causes losses of approximately $100 million to the citrus industry in Säo Paulo. The State of Säo Paulo accounts for about 87 per cent of Brazil's orange production, corresponding to 30 per cent of the world production. The size of the genome was reasonable and attainable with the limited resources.

The laboratories involved were largely from institutions of higher learning, principally universities and research institutions. Teams within the institutions organised themselves and submitted bids to the project managers. The project managers selected the teams on the basis of the information provided by the bidding laboratories. The project did not involve any significant international collaboration. Above all, it was limited to laboratories within Säo Paulo as FAPESP is a Säo Paulo State initiative and not a Federal Government body. ONSA and its partners have contributed three genomes to the public genetic databases: Xanthomonas axonpodis, Xanthomonas campestris and Xylella fastidiosa.

The level of genome sequencing that the Brazilian team has achieved is unparalleled among developing countries. It has attracted the attention of the United States Department of Agriculture (USDA), which contracted the Brazilian team to sequence a train of X. fastidiosa, which afflicts California's grapevines. The team had a steering committee consisting of three experts from Europe and two from Brazil that also served as an advisory body.

The rest of the organisational structure includes a DNA coordinator who was responsible for the distribution of DNA samples (clones) to the sequencing laboratories and two heads for the central sequencing laboratories. The Bioinformatics Centrc coordinated the flow, management and analysis of the data. More significantly, the sequencing laboratories were paid on the basis of the quality of sequences generated. The management of the institute was tailored to encourage generation of high-quality data in the shortest possible time.

The selected laboratories agreed to generate a minimum number of high-quality sequences in a fixed period of time. Laboratories that deposited more good-quality sequences got more money. Further, the representatives of the participating laboratories - 200 participants - met once every four to five weeks in person to review progress and make fresh plans. This was important as daily management was performed via the Internet. The DNA coordinator, the two heads of the central sequencing laboratories and the heads of the bioinformatics centre formed a working executive committee.

These measures partly enabled the project to be completed within time and within budget. In the process they developed their own genome sequencing and analysis methods. For example, they developed a number of bioinformatics tools and a polymerase chain reaction method targeting the amplification of the central portion of expressed genetic material (mRNA) called ORESTES. These tools are now being used in the generation of about one million expressed sequence tags in human genomes cancer research.

Given the success of ONSA, the Ludwig Institute for Cancer Research, an international organisation focusing on cancer, chose Brazil as the location for the express identification of genes relevant to cancer. About $15 million was invested in this project. The project produced over a million human genes sequences whic h contributed to the overall human genome sequencing effort. The project also generated a unique method that primarily amplified the expressed regions of genes. The ONSA project spun off two companies called Alellynx Applied Genomics and Scylla Bioinformatics.

Alellyx Applied Genomics was founded by molecular biologists and bioinformatics experts that spearheaded the establishment and success of ONSA. It employs sophisticated computational software to analyse and identify important genetic information from genome sequence data. These data are then used to isolate the genes and reinsert them in plants to produce recombinant variants with the desired traits.

The company maintains close ties with other advanced institutions involved in genome sequencing, annotation and analysis. It has access to significant genome databases currently available and will generate extra genomic information where the need arises. The initial target of the company is to use advanced computational and biological platforms to analyse and alter the genetic material of soybeans, oranges, grapes, eucalyptus and sugarcane.

With a focused staff establishment of 40 scientists with years of experience in genome analysis and plant molecular biology, the company has strategically placed itself to be a comprehensive, efficient and productive service provider of genetic material for the agricultural sector. Alellyx has strategic alliances with Sun Microsystems and Applied Biosystems.

Scylla Bioinformatics was a dream that became a reality only after the birth of ONSA. The funding of ONSA enabled the initial interests to be tested in a real and highly demanding environment. Having successfully managed the Bioinformatics needs of ONSA, the Sugarcane and the Human Cancer EST projects, the team of professor and students began considering the formation of a private bioinformatics company.

Scylla Bioinformatics aims to provide high-performance software to research centres, industry and other organisations that employ genetic data. The software developed by Scylla was the soul of the distributed network of ONSA. The company aims to provide tools to both the scientific community and private companies that would like to establish virtual networks similar to ONSA.

Scylla also provides bioinformatics integrated software that could be used for reception, clustering, annotation, gene prediction and assembly, as well as tailor-made client specific software for different genomics project needs. The company has formed strategic alliances with Hewlett Packard. Both companies received venture funding from Votorantim Ventures. Votorantim Ventures is a multi-sector venture fund ($300 million) that specifically invests in information and communication technologies (ICTs) and life sciences. The success of the ONSA project has not been limited to the State of Säo Paulo alone.

In 2000, The Federal Government decided to extend the benefits of genomics research to the rest of Brazil. Approximately 25 laboratories distributed across Brazil were selected to sequence Chromobacterium violaceum with a genome size of roughly 4.7 Mb. The organism is abundant in the waters of the Amazon; though opportunistic, it causes a fatal disease.

The organism also produces violacein, a purple colored pigment, with trypanocidal and antibiotic properties as well as molecules that resemble propylene and polyethylene. This project created the Virtual Institute of Genomic Research and involves selected laboratories across Brazil (with an estimated area of about 8.2 million square kilometres).

Similarly, the genome size is larger, almost twofold, than that of Xylella fastidiosa. Therefore, it is important to note that there have been increased genomerelated research activities in Brazil since the initiation of the ONSA projects beyond the State of Säo Paulo. For example, the Virtual Institute of Genome Research has sequenced Mycoplasma synoviae, an organism that infects chicken. In combination with other organisms, it causes fatal respiratory diseases in poultry.

Brazil is the second largest producer of poultry, with export earnings of roughly $1.3 billion. ONSA could be considered to be a centre of excellence with units that are distributed throughout the State. They planned together, and shared human, financial and material resources and management via the web, just like a single institution.

More importantly, they shared work rather than just ideas. This case demonstrates a unique model that could be used by other developing countries in areas such as information and communication technologies and biotechnology. Secondly, virtual research activities are faster to build, and can easily be abandoned, rearranged and recreated to meet new challenges.

They expose a large number of centres and professionals to new techniques and quickly concentrate limited expertise. They render geographical isolation irrelevant, foster collaboration and stimulate quality research activities. The individual participating laboratories expand their research activities on the basis of the lessons learned and use the tools acquired to create and attract new partnerships and sources of funding.

Tissue Culture in the Horticulture Industry

In Kenya and Zambia, over the last two decades, a booming vegetables and cut flowers industry has developed on the shores of Lake Naivasha in Kenya. The population has expanded from 50,000 to 250,000, the majority of whom are women who have been drawn by the opportunities to participate in the production of vitroplants and cuttings which feed the booming industry. This is a highly valued hortic ulture industry that is spurred by simple and efficient biotechnology.

The industry earns Kenya $300 million to $500 million a year. Similarly, Zambia has become the third African producer and exporter of cut flowers, just behind Kenya and Zimbabwe, after a fiveyear growth period. Flower acreage is currently estimated at 135 hectares, most of which is planted with over 40 rose varieties. Zambia exports more than 90% of its flowers to the Netherlands. Tissue culture is also used in banana and cassava production among other crops.

The success of these efforts depends also on the quick movement of produce, and the reduction of market barriers and superfluous regulations. It should also be stressed that workers must obtain a fair return from this booming industry in terms of higher wages and better conditions of life. However, it demonstrates the role that biotechnology is playing in industrial development. Extending these benefits to poor farmers remain a challenge.

Biosidus

In Argentina, the pharmaceutical company Sidus invested in medical biotechnology in the early 1980s and created a subsidiary, Bio Sidus, which started producing its first biotechnology-derived product, a recombinant erythropoietin (EPO). The company recruited five academic scientists and inaugurated its pilot plant to produce EPO, and the other products followed. In 1993, BioSidus became financially autonomous. Approximately, $35 million was invested in Bio Sidus. Today, BioSidus has a tenyear business plan and may become a public company to be quoted on the stock exchange.

The company's organisation has demonstrated, as in the developed countries, that not only good scientists were needed, but also people specialised in marketing, finance, law and regulation. This combination of competences is the key to successfully transforming research results into a marketable product. BioSidus has been awarded several prizes by both the Government and the industry, as well as by the academic circles, to acknowledge its pioneering efforts and success, despite an unfavorable macroeconomic environment.

In 2002, the annual turnover of BioSidus reached $45 million, which can be considered a good performance compared with other biotechnology companies across the world that received funding for a decade without generating profit. About 75 researchers are working in the company and collaboration linkages have been set up with research teams out of the company and even abroad. The current objective is to collaborate with other biotechnology companies in Latin America and the Caribbean in order to manufacture and commercialise products that are useful for the region.

The production of oral vaccines against cholera and typhoid, gene therapy to cure cancer and angiogenesis, and the development of transgenic farm animals to manufacture recombinant biopharmaceuticals (in addition to the production of EPO and interferons). In August 2002, BioSidus announced the birth of a cloned calf, and two months later that of three transgenic calves expressing the gene for the human growth hormone. These were the results of six years of work aimed at producing human proteins in the milk of farm animals.

By 2004, the company may be able to produce the recombinant substance and to transform it into a biopharmaceutical. It is estimated that with a concentration of 1g of protein per litre of milk, a herd of 20 animals would be sufficient to meet the needs of Latin America (the company nevertheless acknowledges that some transgenic goats could produce up to 12g of protein per litre of milk).

Bamboo is an ornamental plant, mainly for garden use in Europe. Besides, two different pathways for valorisation of biotechnological research have emerged: (1) in the tropics bamboo is a very important plant, providing livelihood for over 500 millionpeople and providing housing and shelter for over 1 billion people; (2) "Bamboo for Europe", an EU-funded research project has allowed to develop the potential of bamboo as an agricultural plant in Europe with possible applications in wood industry and as energy crop. So, any biotechnological developments for the improvement of ornamentals will also be useful on a much larger scale.

In Korea amount of horticultural production was not constant yearly, especially spice crops such as hot pepper, garlic, onion and Chinese cabbage etc. Farms have to control themselves not only amount of production but also price.

8

Agriculture and Food Supply Chain

The main features of new agricultural economy are greater concentration/intensity of farms into smaller numbers with large sizes and rising influence of contract farming and the evolution of integrated supply chains linking producers and consumers. Such intensive agriculture relies heavily on electro-mechanical systems, irrigation, agro-chemicals, and basic raw materials that possess new traits such as genetically modified plants and animals.

Under the new agricultural economy, these attributes of intensive farming create new challenges for sustainable production and processing practices that promote a balanced approach to the problems of food quality, safety, and good environmental stewardship.

Modern consumers demand food that is fresh, palatable, nutritious and safe. Furthermore, increasing number of consumers demand functional foods that offer specific health and nutraceutical benefits. With changing lifestyles and rising income in many parts of the world, increasing proportion of foods are prepared and eaten outside the home as convenience and restaurant meals.

These market-pull factors also have major implications on the future of agriculture because they pose considerable challenges for maintaining food safety, environmental protection, and profitability of agriculture. The shift from quantity-oriented agriculture to new emphasis on quality, safety, functionality and sustainability, have placed new demands for the development and adoption of traceable supply chains.

Assuring food traceability from farm to fork, implementing authenticity and diagnostic tests that detect and prevent food safety hazards and preserving the identity and wholesomeness of novel foods, have become essential elements of a quality assured agricultural supply chain management system. Traceability is a preventative strategy in food quality and safety management. However, when hazards or food scares occur, a good traceability system will facilitate timely product recall and determination of liability.

The capability for full trace-back and trace-forward at any stage in the food chain is considered critical to addressing declining consumer confidence and general public concern about the rising incidence of food-related deaths and illnesses, which have been major public health issues in the developed countries.

Advancements in information and communication technology (ICT) for data capture, storage and retrieval, non-destructive testing, and geospatial science and technology provide opportunities and challenges for agricultural engineers to contribute to the development of technological innovations for traceability from farm to fork.

Agricultural Supply Chain Management

The conceptualisation of consumer products and services as supply chains is now a common practice in most industries. From the farming of basic raw materials to delivery of final products to the consumer, each different step in the entire production process is viewed as link in the chain. Supply chain management (SCM), therefore, represents the management of the entire set of production, manufacturing/transformations, distribution and marketing activities by which a consumer is supplied with a desired product. Some analysts refer to this as demand chain management to emphasise the focus on meeting consumer expectations.

The practice of SCM encompasses the disciplines of economics, marketing, logistics and organisational behaviour to study how supply chains are organised and how institutional arrangements influence industry efficiency, competitions and profitability. Technological innovations in information sciences and engineering are increasingly playing vital roles in the vertical integration and coordination of supply chains through the application of hardware and software for measurements, data capture, analysis, storage and transmission.

Two broad principal explanations can be advanced for the increasing interest in agricultural SCM: the industrialisation of agriculture, and the uncertainty associated with variations in product quality and safety. The trend towards vertical coordination of agricultural supply chains (ASC), reduction of government support (subsidies) for agriculture, globalisation and competition among producers, processors and suppliers, explosion in technological progress applicable to the agri-food industry, changing consumer demand and consumption patterns, etc, are some of the factors related to the concentration and industrialisation of agriculture.

The high variability in quality and magnitude that is characteristic of the agricultural environment (both production, handling and processing) and basic food raw materials creates uncertainty in the ability of the industry to assure a consistent supply of good quality and safe products to the consumer. As a result, the transaction cost (information, negotiation and monitoring) associated with market-driven (industrialised) agriculture and the uncertainty of product quality and safety has increased in recent times.

As individual businesses and global markets become highly efficient and cost-effective, competition has therefore shifted from 'between businesses' to within entire value chains. The ever-increasing adoption of SCM principles in agribusiness is perhaps one of the profound and long-term developments in business management relevant to agricultural traceability. Agriculture is inherently a fragmented industry, involving a diverse range of distinct enterprises (farmers, processors, marketers and distributors), and relies on inputs from various sources, often at distinct geographical locations.

For instance, although some food products such as chicken and pork, have been subject to extensive vertical integration within the production and marketing chain, cattle on the other, may be produced on one farm, grown on another, finished on another, and between each stage, transported and sold through open markets, sometimes several times, before being purchased either in the open market or direct, deadweight, for slaughter. With respect to grains and fresh fruit and vegetables, most marketers and processors obtain their supply from diverse sources (farmers, retailers, brokers) in order to meet marketing and production targets.

From the marketing and processing perspectives, SCM is an essential tool for integrating the activities of the various suppliers within the company's operations in order to assure the consistent delivery of qualityassured products and services to the consumer. For the consumer and other stakeholders, SCM focuses on improving the performance of the supply chain through the delivery of guaranteed safe, desirable and good quality food in a costeffective manner.

The increasing transaction costs of intensive agriculture and the need to reduce these costs, lies at the heart of interest in agricultural SCM. Based on experience from a broccoli SCM system in the UK, Grimsdellproposed six fundamental requirements for an efficient supply chain between vegetable growers and the major retail customers: scale of operation, strategic alliances, production flexibility, continuity of supply, quality control, and communication.

The importance of information and communication technology for successful implementation of a traceable supply chain management system. Increasingly, agricultural SCM must be conducted between firms under strategic alliances since most enterprises often represent a link or set of links in the overall supply chain of a product.

Optimising the entire supply chain, therefore, requires a level of information sharing, teamwork, cooperation and collaboration among the participating enterprises. Given this scenario of interdependence, rapid interchange of data on products and activities among companies in a supply chain will reduce transaction costs.

The implementation of supply chain traceability as part of the overall quality management system in agriculture facilitates such data/information exchange by providing a mechanism for recording, storing, analysing and transmitting relevant data on products and activities to designated stakeholders.

Traceability in Food and Agribusiness

The term 'traceability' has become so widely used in recent times in various industriesthat it is timely to examine the concept, particularly in relation to agriculture and food. Agricultural traceability simply refers to the collection, documentation, maintenance, and application of information related to all processes in the supply chain in a manner that provides guarantee to the consumer and other stakeholders on the origin, location and life history of a product as well as assisting in crises management in the event of a safety and quality breach.

With respect to a food product, traceability represents the ability to identify the farm where it was grown and sources of input materials, as well as the ability to conduct full backward and forward tracking to determine the specific location and life history in the supply chain by means of records.

It contributes to the demonstration of the transparency of the supply chain through the use of verifiable records and labeling. Traceability adds value to the overall quality management system by providing the communication linkage for identifying, verifying and isolating sources of noncompliance to agreed standards and customer expectations.

There are six important elements of traceability which put together, constitute an integrated agricultural and food supply chain traceability system:

— *Product traceability*: which determines the physical location of a product at any stage in the supply chain to facilitate logistics and inventory management, product recall and dissemination of information to consumers and other stakeholders.
— *Process traceability*: which ascertains the type and sequence of activities that have affected the product during the growing and postharvest operations (what happened, where, and when). These include interactions between the product and physical/ mechanical, chemical, environmental & atmospheric factors which result in the transformation of the raw material into value-added products; and the absence or presence of contaminants.

— *Genetic traceability*: which determines the genetic constitution of the product. This includes information on the type and origin (source, supplier) of genetically modified organisms/materials or ingredients as well as information on planting materials (such seeds, stem cuttings, tuber, sperm, embryo) used to create the raw product.

— *Inputs traceability* : which determines type and origin (source, supplier) of inputs such as fertilizer, chemical sprays, irrigation water, livestock, feed, and the presence of additives and chemicals used for the preservation and/or transformation of the basic raw food material into processed (reconstituted or new) food products.

— *Disease and pest traceability*: which traces the epidemiology of pests, and biotic hazards such as bacteria, viruses and other emerging pathogens that may contaminate food and other ingested biological products derived from agricultural raw materials.

— *Measurement traceability*: which relates individual measurement results through an unbroken chain of calibrations to accepted reference standards. To achieve this, measuring and test equipment and measurement standards are calibrated utilizing a reference standard whose calibration is certified as being traceable to a national or international standard.

The other aspect of measurement traceability relates to the property of the measurements (data and calculations) generated throughout the supply chain and their relationship to the requirements for quality. By focusing on the quality of measurements (rather than on a property of an instrument, it is possible to assure that the measurements are indeed adequate for the intended use. To achieve this, each measured data must specify the environmental, operator, and geospatial and temporal factors, which are not related to the instrument but impact on the quality of the data.

In implementing a new traceability system or studying an existing one as part of routine quality management system or in the event of food safety and quality alert, these basic aspects must be addressed in order generate sufficient data to adequately evaluate the type, origin and location of the source of safety concern to enable corrective actions to be taken.Traceability is an information-based proactive strategy to food quality and safety management. It is a complimentary tool to other quality management programmes such as Hazard Analysis and Critical Control Points (HACCP) systems.

A key strength of traceability chain management is that it facilitates the identification and isolation of hazards and implementation of effective corrective actions in the event of an incident. Thus, like point inspection and product testing, traceability by itself cannot introduce safety into the food process or handling process. When considered in isolation of other quality management systems, its traceability is not a sufficient condition to satisfy the safety requirements of the food chain.

However, its strength lies in preventing the incidence of food safety hazards, and reducing the enormity and impact of such incidents when they occur by facilitating the identification of product(s) and/or batches affected, specifying what occurred, when and where it occurred in the supply chain, and identifying who is responsible. The benefits of integrating traceability into the overall quality agricultural management system are numerous, ranging from improvements in product quality and safety management, crises management in the event of a safety alert, and strengthening overall agribusiness coordination.

With heightening public scrutiny of the food supply chain and agriculture, many national and regional new food quality regulatory directives and laws have been enacted, leaving agriculture and food industries with little option but to implement traceability systems as part of the overall food safety and quality management programme.

As agriculture continues to experience declining terms of trade and competition by other more financially lucrative industries, there are good reasons to believe that the concern about traceability will continue in global food trade. The search for cost-effective technological innovations for implementing accurate and reliable traceability systems is therefore an important challenge facing agriculture in the new globalised economy.

Demand for Traceability

The demand for traceability in agriculture has significantly expanded in the last few years with increasing incidence of food-related safety hazards and scares (such as footh-andmouth disease, mad cow disease, microbial contamination of fresh produce, dioxin in poultry). The appearance of genetically modified organisms (GMOs) and the need for identity preservation of GMO and non-GMO agricultural chains has further exacerbated declining consumer confidence on food safety and the increasing concern over potential negative impacts of agriculture on the environment and ecological diversity.

Closely related to these factors is the introduction of new food safety legislations that have placed responsibilities on producers, processors, caterers and other handlers in the supply chain to ensure food safety. The main tool that these operators will use to defend any subsequent liability will be the demonstration of "due diligence", which must demonstrate that every precaution has been taken to prevent contamination and subsequent food safety hazards.

More than before, an essential feature of food quality management system is that finished products can be traced back to their raw material and thus to their original producer and previous handlers in the chain. In the same way, forward traceability is also essential to guarantee the location of products and facilitate their recall when safety and quality standards have been breached.

Technological Implications

Modern agriculture is highly knowledge-intensive and increasingly information-driven. With declining terms of trade affecting agriculture vis-à-vis other industries, technological innovations are necessary to reduce transaction costs and facilitate the production and consistent supply of top quality, safe and traceable products to meet consumer demands. Farmers, processors, marketers, handlers, consumers, governments and the general public have stakeholder interests in the safety and cost-effectiveness of global agri-food supply chains.

To implement traceable agricultural supply chains, technological innovations are needed for product identification, process and environmental characterisation, information capture, analysis, storage and transmission, as well as overall system integration. These technologies include hardware (such as measuring equipment, identification tags and labels) and software (computer programmes and information systems).

Product identification technology: A major feature of any traceable supply chain is the ability to trace-back the history and consistently track the physical location of the products in the overall supply chain. To achieve these, accurate labeling is essential. The simplest technology to achieve this is to attach a tag to the basic raw material (e.g. the animal, plant or its constituent parts) and to transfer that data on the tag to the bar code of the food product.

In the livestock industry, the ear-tag is commonly used and often contains a series of numbers and/ or alphabets, which codify specific information such as the breed, date of birth, farm, vaccinations, etc. The use of computers and other information technologies have spurred the development of electronic identification (EID) systems, which include electronic tags with chips and handheld scanners for reading, storing and transmitting the data to PCs for analysis and long-term storage.

An important attribute of tags is that the materials must be resistant to rough handling and inclement weather. Advancements in material science have led the development of tags that are resistant to tear and wear and which can withstand harsh environmental conditions. Innovations in geospatial science and technology such as radio frequency technology and mobile tracking devices have the potential for collecting and transmitting data from tags to distant locations for storage and analysis.

The simplest label on a food product often stores such information as the name of the product, batch/lot number, and price. Depending on the market sophistication and demand for information, other data that can be loaded are product origin, handling and storage conditions.

Quality and safety measurement technology: Ultimately, the success of traceability is measured by the consistent delivery of products to meet the expectations of the consumer and other stakeholders, as well the ability to ascertain the location of each product unit for effective recall in the event of a quality or safety breach.

This requires accurate information on the maturity and quality attributes and safety status of the product, which must be measured and analysed using appropriate instrument and procedures. Product specifications such as size (mass, dimensions), firmness (crispness, crunchiness) soluble solids, acidity, flavour, etc, are some of the physical, mechanical and chemical properties that may require measurement.

Laboratory and on-line instruments such as the penetrometer, firmometer, twist tester, Instron machine, and Kiwifirm can be used to measure the firmness of fleshy products. Nondestructive tests based on force sensing, infrared and magnetic resonance imaging can also be used to measure firmness and other internal quality attributes. They can also be used to assess the presence of hazardous physical objects inside products.

Models of these instruments are commercially available from various manufacturers and are currently used in industry and research laboratories. Procedures and equipment are also available for chemical analysis of product samples, equipment surfaces and the air to ascertain the presence of hazardous microbial contaminants.

Genetic analysis technology: The need to preserve the identity of product supply chains and the demand for genetic traceability have led to the development of procedures and measurement devices for the analysis of the genetic constitutions and contamination of foods and other biological products. DNA tests based on real-time PCR have been developed to detect and quantify GMOs and other transgenic materials. Gieseprovides a good summary of several laboratory exhibits at a recent Institute of Food Technology Meeting & Food Expo for measuring product traceability and safety.

Environmental monitoring technology: Environmental conditions such as temperature and relative humidity, atmospheric composition of the air, including pollutants, impact on the quality stability and safety of food products. Instrumented environmental recording devices (such as gas analysers and biosensors) for monitoring these parameters are available and can be coupled to control systems.

Developments in geospatial science and technology: The integration of geographic information systems (GIS), remote sensing (RS) and global positioning systems (GPS)offers considerable opportunity for site-specific agriculture and the derivation of data related to the agriculture products, on the farm and through the subsequent handling operations.

Put together, these technologies enable data to be remotely collected on individual animals, plants, and blocks on a farm, which can be processed, transmitted and presented as visual spatial information on the bio-physical properties of the block, yield and selected product attributes. With respect to traceability, a vital feature of these technologies is the possibility to map the geospatial variability of selected attributes such as yield, product quality, animal movement, and disease epidemiology.

Software technology for traceability system integration: An effective traceability chain is an integrated system made of distinct components involving data collection using standards measurement procedures, the analysis, storage and transmission of the recorded information, and full backward and forward control system that permits the tracking of the product history.

Such an integration of technologies relies on the application of appropriate information and computer system, and which links the traceability chain to a central database at the company, national or international level. Many researchers and commercial companies have reported the development and availability of full traceability systems that have been adopted in industry.

Calder and Marrdescribed a trial traceability initiative based on electronic identification (EID), which enabled full traceability of farm animals. Harveyrecently reported an electronic data identification tag (EDIT) that includes a transponder as part of a typical ear tag, and a handheld, battery-powered electronic reader or terminal, which provides farmers with a fully integrated traceable livestock management tool.

The electronic ear tag has the capacity to record all the details of an animal's life from its genetic and birth details, to treatment and productions. Many countries such as Canada and New Zealand have implemented national cattle identification programmes using various technologies that are based on the above principles. Examples of commercially available traceability software programmesthat have been reported include QualTrace, EQM (Enterprise Quality Management), and Food Trak

Traceability has become a critical element in global agri-food business. Shifting patterns in consumer demand for sustainable agricultural practices (including environmental and animal welfare concerns) and ever-increasing demand for consistent supply of a range of top quality and safe products assure a future for traceability as an important element of the overall quality assurance system. Traceability presents both challenges and opportunities for educators, research and development practitioners and agribusinesses.

Developing cost-effective traceability technologies for both large and smallscale farmers and postharvest enterprises should receive priority attention by engineers and other science and technology experts working in agriculture and other biological industries. Farmers and food product handlers need training on the principles and procedures of traceability. With increasing concentration of global supply chains and enactment of new regulations on traceability in agriculture, technology transfer and rural development projects designed to improve the marketorientation and quality of products from small-scale farmers in developing countries must include appropriate tools to facilitate the traceability of their products and processes.

Several technologies already exist, particularly in the livestock industry, for implementing traceable supply chains. Future innovations in DNA finger-printing, nanotechnology for miniature-machines, and retinal imaging and their integration into plant and livestock industries have considerable potentials for improving the speed and precision of traceability in knowledge-based agriculture.

In view of the prospects for litigation arising from product liabilities, the development of relevant measurement and calibration standards related to product traceability demands the attentions of researchers and policy agencies dealing with food quality and safety.

9

Comparative Analysis of Fertilisers

Fertilisers enter the pond in either soluble or particulate form, and deliver soluble P, N, and/or C for algal uptake and growth (Figure 1).

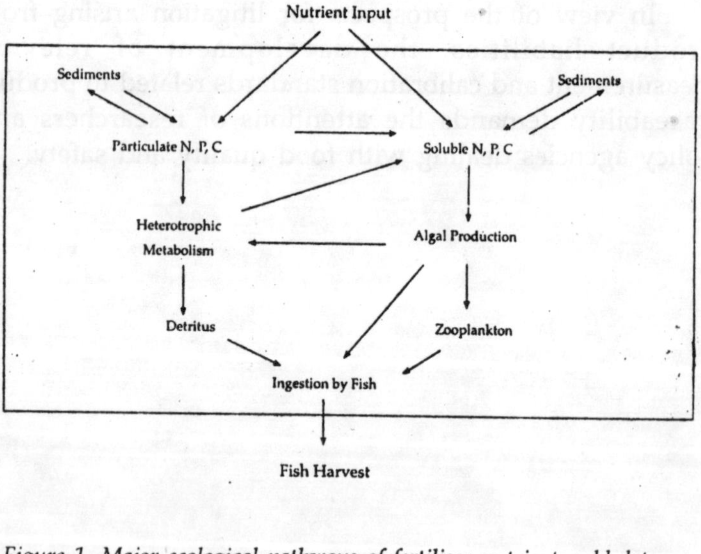

Figure 1. Major ecological pathways of fertiliser nutrients added to a culture pond.

Fertilisers analysed can be divided into organic and chemical varieties. Organic fertilisers are either animal manures or plant compostings, while chemical fertilisers include nitrogen-phosphorus-potassium (N-P-K) fertilisers, triple superphosphate (TSP), sodium nitrate, and urea.

Analytical Factors

Fertilisation Goals

The goal of pond fertilisation can be viewed simply as maximising the causal link between fertiliser inputs and ultimate yields of culture organisms at harvest, while minimising economic and environmental costs. This general goal encompasses the four specific objectives in fertilisation theory, namely:

— To increase natural food production by stimulating algal productivity;

— To optimise nutrient utilisation efficiency;

— To optimise cost efficiency; and

— To maintain a favorable growth environment for culture species.

Too often a fertilisation strategy is evaluated solely on its ability to increase yields of culture organisms. If you get bigger fish, it works; if you did not get bigger fish, it did not work. This is rarely the most cost-efficient approach. For example, researchers have put high inputs of N and P into shallow, turbid ponds, and then made general recommendations that high nutrient input rates are not beneficial because fish did not grow any better.

That may be true for shallow, turbid ponds. By appreciating the fact that light limitation was preventing further algal productivity, however, a more useful recommendation would have been to deepen the ponds (or raise the banks) so the turbidity would settle out of the surface waters. This approach increases light availability and allows the algal community to benefit from higher rates of nutrient input. Only by understanding dynamic processes in the pond and why a particular fertilisation strategy did or did not give expected results can more accurate costs and benefits be evaluated.

The aim here is to highlight impacts on pond ecology and additional costs of mitigating resulting problems which should be considered by farmers and/or extension workers when making fertilisation strategy decisions, and by researchers when making fertilisation recommendations to rural farmers. Just getting bigger fish or higher yields is not enough if actual economic and environmental costs are too great. It is critically important that farmers appreciate the broad economic and ecological implications of choosing one fertilisation strategy over another.

For technical assistance in making more quantitative, in-depth economic analyses of aquaculture systems, the reader should consult more comprehensive sources. Economic factors considered in the following comparative analysis of fertilisers include:

— Market cost of biologically available N, P, and C (e.g., on-farm production versus off-farm purchase);

— Transport and labor costs, including lost opportunity costs and integration of aquaculture with other farm activities;

— The need for additional capital expenses, such as for pondwater aeration or mixing;

— Market value of culture organisms (e.g., social acceptance/ marketability of manure-raised fish); and

— Impacts on pond ecology, including water quality (e.g., dissolved oxygen, ammonia, turbidity, diseases) and the filling of ponds with organic matter.

Getting the most "economical" fertiliser is site-specific. Fertiliser requirements should be evaluated on a farm-by-farm basis, and can even be done on individual ponds.

To illustrate, three ponds located side by side would require totally different approaches to fertilisation if we assume they have different depths of 0.7 m, 1.2 m, and 3 m. Individual pond differences in thermal stratification characteristics and ecological impacts of resuspended bottom sediments greatly influence each pond's response to identical fertilisations.

Before adopting any country-wide or regional fertiliser recommendation, therefore, its ecological foundation/assumptions should be examined critically to ensure that the suggested fertilisation strategy is appropriate for each pond or farming system in question. This goal of fertilisation is a rather broad topic, and is covered more extensively elsewhere. The focus here is limited to two areas of pond ecology where the choice of fertilisers can have a substantial influence on the health and welfare of culture organisms:

— the need to maintain sufficient dissolved oxygen in the pond water; and

— the need to minimise the presence of toxic, unionised ammonia concentrations.

Dissolved Oxygen

Most aquatic organisms require dissolved oxygen (DO) for metabolism and growth, and without oxygen these organisms will quickly die. Since air is about 21% oxygen ($\approx$ 300 mg l^{-1} of air), oxygen depletion is rarely an issue for terrestrial organisms unless they are suffocated. Depletion of dissolved oxygen in aquatic environments, however, is a serious, ever-present concern.

The solubility of oxygen in water is less than 1% of its solubility in air, and decreases further with increasing water temperatures. Saturated DO concentrations in water at 1 atmosphere of pressure range from about 9.1 mg l^{-1}

at 20°C to about 7.0 mg l^{-1} at 35°C. These values show that there is not a lot of DO in the water when compared to O_2 concentrations in air.

Different aquatic organisms require different minimal levels of DO in order to survive, or at least not be detrimentally affected. For example, freshwater prawns may suffer when DO falls below 3 to 4 mg l^{-1}, whereas Nile tilapia can easily tolerate DO concentrations below 1 mg l^{-1} for short periods of time. Fertilisation should be conducted in a way to maintain DO levels above minimal acceptable values for the target culture species.

Oxygen production during daytime photosynthetic activity and oxygen uptake for biological respiration and biochemical oxidative reactions (e.g., nitrification) create the typical diel DO. Minimum DO concentrations are normally found at pre-dawn, just before photosynthetic activity begins with sunrise. Maximum DO concentrations can reach over 30 mg l^{-1} in very productive ponds in the late afternoon. Supersaturation of DO can exist because of water pressure, but eventually some DO bubbles out when subsurface water is brought to the surface.

Even in the most productive ponds, pre-dawn DOs will usually remain about 3 mg l^{-1} as long as the organic matter in the pond is produced only within the pond. In other words, very green ponds which do not receive any additional organic matter (e.g., manures or feeds) should not experience early morning DO levels below 3 mg l^{-1}. This assumes that ponds are shallow enough to experience nighttime destratification and wholepond mixing.

As more organics are added to the pond, the oxygen demand increases and nighttime DO concentrations decrease. If too much organic matter is added to the pond, DO can be depleted, causing potentially lethal, anoxic conditions to develop. It is possible to add DO to pond water through appropriate pond management, but with additional corresponding costs and risks.

For example, mechanical aerators, such as paddlewheels, propelleraspirator pumps, and vertical-pump aerators can be particularly effective in adding DO. Oxygen diffusion from air into calm water is quite slow, but can be accelerated substantially by agitating the water's surface.

Included in the farmer's decision whether or not to use organic fertilisers, therefore, are considerations of the financial and labor costs associated with the purchase, operation (including energy costs), and maintenance of these devices. Furthermore, ponds which require aerators to maintain minimal DO concentrations above lethal levels put the farmer at a greater risk, particularly if fish survival depends upon not having any mechanical or power failures.

Un-ionized Ammonia Toxicity

The second major environmental concern is with un-ionized ammonia toxicity. Remember that ammonia occurs in water in two soluble forms, namely ionized (NH_4^+) and un-ionized (NH_3). To facilitate discussion, the equilibrium reaction introduced in Equation 1 is repeated:

$$NH_4^+ + OH^- \rightleftarrows NH_4OH \rightleftarrows NH_3 + H_2O \quad [1]$$

This transformation between the ionized and un-ionized forms of ammonia is a function primarily of water pH and, to a lesser degree, water temperature. The equation moves to the right (i.e., the transformation of the ionized to the un-ionized form) with increasing pHs and, to a much lesser extent, with increasing temperatures. Table 1 illustrates this relationship, showing the different percentages of unionized ammonia in the total ammonia pool with variable pH at 25°C.

Table 1. Relationship between pH and relative equilibriumpH NH_4 concentrations of total dissolved ammonia.

pH	NH_4^+	NH_3
5	100.0	0.0
6	99.9	0.1
7	99.4	0.6
8	94.7	5.3
9	64.2	35.8
10	15.1	84.9
11	0.8	99.2

This transformation has importance in aquaculture systems because, unlike the ionized form, the un-ionized form can be highly toxic to culture organisms. However, a word of caution is necessary before assuming that reduced fish growth may be related to high ammonia levels in the pond. For example, in a fertilisation study using fixed weekly inputs of N and P, tilapia net fish yields (NFY) in some ponds were comparatively low and total ammonia concentrations were several milligrams per liter.

Upon further examination, it was shown that lower fish yields corresponded directly to lower natural food production as indicated by lower algal net productivities (NP). Low NPs were related to both DIC and turbidity-induced light limitations, conditions which caused the accumulation of unused ammonia-N in the pond water (Figure 2). Therefore, there was no direct causal connection between observed high ammonia concentrations and low NFYs.

The above example further illustrates how an ecological understanding of pond dynamics can provide direction for a more efficient fertilisation strategy. Rather than reduced N inputs, these ponds needed more DIC and reduced inorganic turbidity to stimulate algal productivity.

Fertilisation with DIC would allow the algae to utilise surplus N, thus reducing ammonia concentrations while at the same time increasing both NP and NFY.

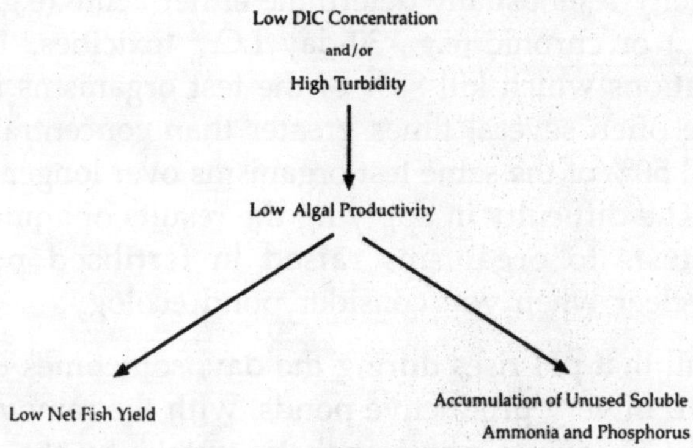

Figure 2. Impacts on algal and fish productivity when a production pond fertilised with N and P is limited by the availability of dissolved inorganic carbon (DIC) and/or availability of light by high inorganic turbidity.

Further evidence that un-ionised ammonia may not be as serious a problem as generally attributed, at least for tilapia culture, was shown in a statistical residual analysis of the relationship between NFY and NP from over 60 fertilised ponds (Knud-Hansen, unpublished data). The analysis found no lethal or sublethal effects on growth when Nile tilapia were exposed to afternoon, un-ionized ammonia concentrations of between 2 and 3 mg l^{-1}.

These results conflict with published 48-hour LC_{50}s of around 2.1 to 2.4 mg l^{-1} un-ionized ammonia for tilapia based on static or continual-exposure toxic bioassays, although Daud et al. reported higher values for red tilapia fry. To understand why ammonia toxicity may not be as great a problem in fertilised culture ponds, it is important to review toxicity testing methodology. Ammonia toxicity for different culture species is typically determined in laboratory experiments, subjecting test organisms to continual exposure to a gradient of un-ionized ammonia concentrations. The concentration which is lethal to 50% of the test organisms is called the LC_{50}.

Toxicity tests usually determine either acute (e.g., 24-hour LC_{50}) or chronic (e.g., 30-day LC_{50}) toxicities. Toxin concentrations which kill 50% of the test organisms in 24 hours are often several times greater than concentrations which kill 50% of the same test organisms over longer time periods. The difficulty in applying the results of ammonia toxicity tests to organisms raised in fertilised ponds becomes clear when you consider pond ecology.

Recall that pH rises during the day, sometimes up to a pH of 10 in very productive ponds, with the removal of dissolved CO_2 due to photosynthetic uptake by the algal community. Excess ammonia not taken up by algae is converted to the un-ionized form in greater percentages as the pH rises (Table 1). But this high pH is a temporary condition, and after sunset the pH begins to drop back to about 7 in most ponds. In highly productive ponds, high pHs may be found only in the upper 20 to 30 cm since photosynthetic activity is light-limited in subsurface waters due to algal self-shading.

From a fish's perspective, un-ionized ammonia concentrations may be elevated in the surface waters for only a few hours during the day. The fish have most of the day and all of the night to recover from any sublethal effects and to acclimate to sublethal concentrations. Moreover, fish can swim to refuge areas (e.g., deeper water) when ammonia concentrations in surface waters are high. This recovery period is not incorporated in ammonia toxicity tests, and relying on 72- or 96-hour LC_{50}s based on continual exposures may overestimate the importance of ammonia toxicity in fertilised ponds.

A 4-hour LC_{50} may more realistically reflect un-ionized ammonia concentrations harmful to fish raised in outdoor, fertilised ponds. However, the fact that un-ionized ammonia may be of lesser importance in fertilised ponds does not mean that the farmer should not reduce any potential risks. The two management options are to keep afternoon pHs lower, or to keep afternoon total ammonia concentrations lower. Increasing the buffering capacity by increasing alkalinity (e.g., with addition of lime) will reduce pH fluctuations.

However, a more cost-efficient approach would be to fertilise with nitrogen more efficiently. In nitrogen-limited ponds, the algae utilize essentially all the dissolved inorganic N available to them. Ideally, however, the pond should be so productive that light limitation from algal self-shading controls algal productivity. Under these conditions it is important to supply only as much N as the algal community requires, which is the simplest way to avoid the possibility of un-ionized ammonia toxicity in culture ponds.

Before examining the fertilisers themselves, it is useful to reflect on what we have already learned, and what we need to know to best evaluate a given fertiliser's utility in a culture pond. We know that:

— Algal productivity is generally limited by availability of P, N, C, and/or light;

— Fertilisers provide soluble P, N, and/or C for algal uptake;

— It is not desirable to have high inorganic turbidity, critically low DO concentrations, or high un-ionized ammonia concentrations; and

— The relative utility of a fertiliser is based on comparative economics, including potentially negative impacts on pond ecology.

With this in mind, it is time to evaluate the more common fertilisers used in warmwater aquaculture. The three categories examined are animal-based organic fertilisers, plant-based organic fertilisers, and chemical fertilisers.

Manures of Animal

Animal manures have a long history in aquaculture as sources of soluble P, N, and C for algal growth and natural food production, and as sources of particulate organic matter for rotifer production. Here does not examine every possible source of animal manure suitable for aquaculture, but instead focuses on how different types of manure provide soluble algal nutrients, impact pond ecology, and satisfy the farmer's need for cost-efficient fertilisation.

For purposes of discussion, animal manures are categorized as either poultry (e.g., chickens and ducks) or mammal. Manures from mammals are either from ruminants (e.g., cows and buffaloes) or non-ruminants (e.g., pigs and rabbits).

Nutrient Availability

The first consideration is determining how well each fertiliser provides soluble P, N, and C for algal uptake. It must be appreciated that only a fraction of the P, N, and C in the manure will become available for algal growth. Most of the nutrient release occurs within a few days of adding the manure to the pond, primarily through leaching and the breakdown of soluble organic molecules.

A certain percentage of manure-P, -N, and -C will remain bound in particulate matter, and will eventually be buried in pond sediments. Economic analyses of fertilisers must reflect the cost per molecule of only the nutrient pool made available for algae, without including the percentage of nutrients lost to the sediments and never made available to algae. Manure nutrient concentrations, and the percentage of manure-P, -N, and -C which becomes available for algal uptake, depend primarily on the animal's diet, whether the manure is liquid or solid, and the age and storage conditions of the manure.

First, the source-animal's diet is important because what comes out of an animal is directly influenced by what it consumed. Animals fed highprotein diets typically have manures richer in N and P than manure from similar animals who rely on scavenging for their sustenance. Appreciating the relationship between animal feed quality and resultant manure quality helps the farmer reduce the risk of overfertilising the ponds, which could result in ammonia toxicity and pond water deoxygenation.

Second, nutrient availability is related to the consistency or form of the manure. Liquid excretions (e.g., urine) already contain algal nutrients in soluble forms, and more nutrients will be released with the decomposition of soluble organic matter. Algal nutrients are generally more tightly bound in particulate manure, and so a smaller percentage of the total manure-N, -P, and -C in solid wastes will become available for algal uptake. The third factor affecting the percentage of available nutrients is the age and storage conditions of the manure. Fresh manure contains more nutrients than manure which has been stored.

During manure storage, aerobic decomposition results in the release of CO_2 and ammonia, reducing both the total amount and percentage of N and C available to algae when the manure is eventually put in the pond. The loss of CO_2 may be compensated somewhat by lime (calcium carbonate), which is often added to stored manures to reduce noxious odors. If the stored manure is not protected from the weather, leaching of soluble nutrients will further decrease the fertilisation value of the manure. Table 2 compares nutrient concentrations of several typical animal manures.

Table 2. Representative availabilities of nitrogen (N) and phosphorus (P) in various animal manures.

Animal	% Moisture	% N	% P
Poultry littera	28	2.8	1.2
Chickenb(bagged)	38	2.0	3.0
Duckc(fresh)	82	3.7	1.7
Buffalod (fresh)	77	1.7	0.1
Dairy cattlea (fresh)	86	0.5	0.1
Swinea(fresh)	89	0.6	0.2
Sheepe (fresh)	77	1.4	0.2

There are many benefits of using animal manures for pond fertilisation. Manure can be a good source of CO_2, which may be needed in rain-fed or other ponds with low alkalinities. Although manures do not increase alkalinity unless lime was added during storage, the CO_2 released during decomposition will be available for algal uptake. Manure can supply soluble N and P for algal utilization, and provide a substrate for zooplankton production. Manure additions may also help clarify clay turbidity in pond water.

Furthermore, a layer of organic matter on the pond bottom can help reduce the rate of P adsorption to pond sediments and may reduce seepage of pond water. The final benefit of manure may be its availability and easy production on the farm. It may be argued that manure can be consumed directly, thus providing an additional benefit for culture organisms.

Utilization of particulate chicken manure by Nile tilapia in ponds fertilised only with chicken manure. In highly productive ponds receiving primarily chemical fertilisers, however, there was no significant benefit to tilapia when the fertilisation regime was supplemented with chicken manure. The nutritive value of animal manure is poor when compared to living algae, zooplankton, and algal-based detrital aggregates. In fact, suspended particles of manure may actually impair fish productivity by diminishing the overall nutritive quality of the filterable organic matter in the pond.

Furthermore, adding additional manure for purposes of direct consumption may unnecessarily degrade the quality of pond water. Manures can be beneficial in providing algal nutrients to stimulate primary productivity, and decisions to add manure should be based on its utility for producing natural foods.

Environmental Impacts

Although manures may represent a readily available source of algal nutrients, they contain other elements which can have serious and undesirable impacts on pond ecology. First, manures contain organic matter. Other than the possibility that culture organisms may in fact consume particles of manure, the main problem with introducing organic matter to a pond is the depletion of DO required for its utilization and decomposition. As more manure is added to a pond, pre-dawn DO levels can reach 0 mg l^{-1}, causing severe stress or mortality of the culture organisms.

The mixing of anoxic bottom water back into the water column can potentially increase ammonia and hydrogen sulfide concentrations to toxic levels. Pond filling is an additional consideration associated with fertilising with particulate manures. The depth of accumulated organic solids increases over time, thus reducing the effective volume of water available for culture organisms.

Organic accumulation is less of a problem in tropical ponds where water temperatures remain relatively warm all year, and when bottom sediments are resuspended into oxygenated waters above. Nevertheless, heavily manured ponds may periodically require the costly shoveling-out of bottom sediments. Manures produced by ruminants, such as cows and buffaloes, release dissolved organic compounds which can also degrade the pond environment. The plant material consumed by these animals contains complex organic molecules, which are passed into the manure.

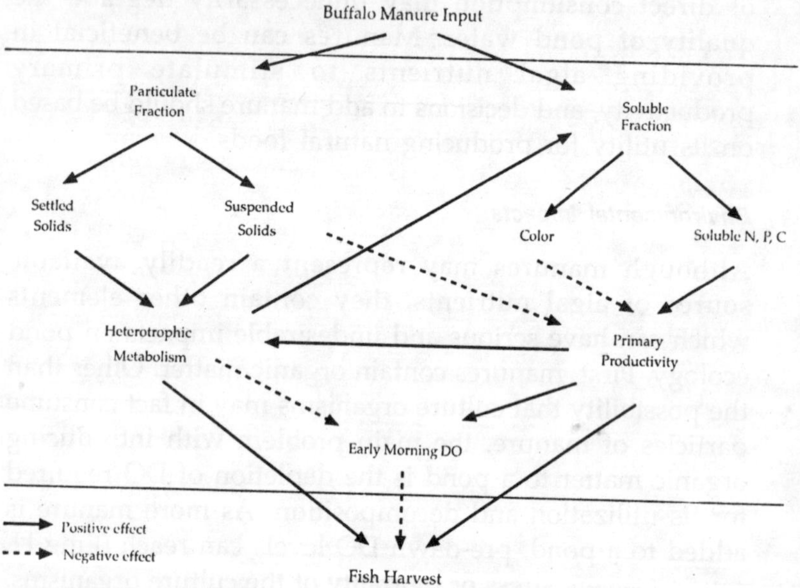

Figure 3. The positive and negative ecological impacts with the addition of buffalo manure in a culture pond.

These soluble organic molecules impart a dark color to the water, thereby reducing the amount of light available for algal photosynthesis. The more these manures are used, the lower the potential net algal productivity, and the lower the resultant net fish yields. Increasing water color and decreasing dissolved oxygen with increasing rates of buffalo manure fertilisation.

Figure 3 illustrates the positive and negative effects on pond ecology when putting buffalo manure in a pond. These dissolved organic molecules will eventually decompose over time, but until then they represent an undesirable ecological impact on natural food production.

Farm Costs

Although the animals do not charge anything for giving it, from the farmer's perspective manure is not free. Manure available for aquaculture is either purchased off-farm, or produced on-farm. If purchased off-farm, associated costs include the actual purchase price, transportation costs to the farm, labor for hauling and shoveling the manure into ponds, and lost opportunity costs.

Here, lost opportunity costs refer to opportunities for economic improvement the farmer no longer has because of time and financial expenses devoted to obtaining and putting fertilisers in his/her ponds. These costs vary with each farmer and locale, but it is incorrect to assume that the farmer has nothing better to do than shovel manure-it should be up to the farmer to make that determination. Additional farm costs may include the purchase, maintenance, and operation of mechanical aerators, as well as costs associated with removing accumulated bottom sediments from the pond.

When buying manure off-farm, purchase-price comparisons with other fertilisers are best accomplished by calculating the actual cost of the available algal nutrients released by the fertiliser. A cost comparison between chicken manure and chemical fertilisers (TSP and urea) in Thailand showed that chicken manure was over seven times more expensive as a source of available N than urea, and over four times more expensive a source of available P than TSP.

This result was surprising in that a 50-kg bag of chicken manure cost only 20 baht (US$1 = 25 baht at the time), while 50-kg bags of urea and TSP cost 240 baht and 450 baht, respectively. But because of the concentrated nature of urea and TSP, 1 kg of urea and 1 kg TSP together provided an amount of available N and P equivalent to about 100 kg of chicken manure.

This ratio will vary depending upon the quality of manure and the solubility of algal nutrients contained therein. For example, Nath and Lannan found 60 to 80% of the total N and P in chicken manure was available. Such a price comparison is only one aspect of the economic analysis, however, and the other relevant factors of associated farm expenses (e.g., manual labor, time, and impacts on pond ecology) must also be considered when choosing the most appropriate fertiliser.

The easiest way to eliminate the purchase price of manure is to produce it on-farm. Using the output of one farming subsystem as input for another is a common component of an integrated farming system. In this case, animals can be raised adjacent to aquaculture ponds so that waste products can be conveniently washed into the pond. If waste disposal is an economic/labor cost to the farmer, then integrated aquaculture can provide an additional benefit. Integrated farming with poultry can be as simple as putting chicken coops over a pond, or allowing ducks to spend the day swimming on the pond.

In the former, spilled feed will also contribute algal nutrients through decomposition as well as through direct ingestion by fish and subsequent excretion. With ducks, it is best to have duck houses also over water (e.g., on stilts) to prevent the ducks from eroding the pond banks with daily traffic. Pond bank erosion will increase inorganic turbidity, and thus reduce available light for algal productivity. Duck houses should not inhibit water movement underneath, and ponds should be large enough so that duck houses do not cause significant shading of pond water. When evaluating the economics of an integrated system, the focus should be on the individual subsystems (e.g., pond, crop, poultry, pig).

If all subsystems are showing a profit, then using integrated farming to supply pond fertilisers can improve farm profitability. For example, if raising pigs is profitable, then the corresponding pig manure available for pond fertilisation is an additional value of the pig operation. If the pig subsystem loses its profitability due to increasing production costs or a significant drop in the market value of pork, however, it makes little economic sense to raise pigs just to fertilise ponds if the net loss from the pig production is greater than the cost of other available fertilisers.

Even if all subsystems are profitable, an itemized review of total farm expenses may indicate that some other available sources of pond fertilisers are more economical in the long term.

Green Manures

The use of terrestrial and aquatic macrophytes for aquaculture is a very broad topic, which includes direct feeding on freshly cut plant material and the composting of plants both in and outside the pond. Composted plant materials provide a pond with decomposed particulate matter, and release soluble algal nutrients and dissolved organic matter during decomposition.

The first stage of plant decomposition is characterized by the leaching of soluble organic compounds during the first several days after submersion into pond water. These leachates are rapidly decomposed and release ammonia, soluble P, CO_2, and other by-products . Microbial oxygen demand remains high while the easily decomposable materials are broken down. This first stage is followed by the considerably slower decomposition of remaining particulate plant matter and more complex dissolved organics. Included in the latter group are tannins and other dissolved organic materials, which impart a dark and lasting color to the water .

Plants utilized as green manures include both terrestrial grasses and nitrogen-fixing legumes, as well as rooted aquatic macrophytes. The best plants are low in fiber (e.g., aquatic macrophytes) so decomposition is faster, and high in soluble algal nutrients (e.g., nitrogen-rich legumes). Plants which have been chopped into pieces decompose relatively faster than unchopped plants because of increased surface area for microbial attachment and decomposition.

Plant material may be added directly to the pond, or put into separate composting areas. Composting may be conducted in the pond in a number of different ways. In China, farmers pile plant biomass in 150-kg heaps along the pond banks. After three to four days the heaps begin to leach into the water, turning the pond a greenish-brown color. Heaps are spread several times to distribute organic matter throughout the pond. Within ten days, the partially decomposed plant material is removed from the pond and incorporated into the soil for terrestrial farming.

Another technique involves using bins in the corner of a pond. Plant matter in the bin leaches dissolved organic matter into the pond, but keeps the particulate matter out of the culture area. Composting bins may also be located outside the pond with the leachate either filtered out or washed directly into the pond.

Nutrient Availability

Nutrient availability from green manures depends on a number of factors. Different plants have different amounts of N and P, and the concentrations of N and P vary with the different parts of the plant . For example, leaf biomass generally has greater concentrations of leachable N and P than woody stems or roots. Plants fertilised with N and P will have greater concentrations in their biomass than similar plants grown with reduced N and P availability. Table 3 provides N and P concentrations for several different plants.

Table 3. Representative availabilities of nitrogen (N) and phosphorus (P) in various plants.

Plant	% Moisture	% N	% P
Alfalfa (dry)	10	2.4	0.2
Alfalfa (fresh)	76	0.7	0.1
Mixed grass (dry)	11	1.1	0.2
Mixed grass (fresh)	69	0.8	0.1
Rice hulls	8	0.5	0.1
Rice straw	7	0.6	0.1
Sugar cane leaves	74	0.2	0.1
Cottonseed meal	7	6.9	1.1
Soybean meal	10	7.3	0.6

Whether plants decompose in composting bins or directly in the pond, the decomposition processes are the same and the amount of nutrients eventually released from the plants should be reasonably similar. There are two exceptions to this general assumption. First, placing compost bins outside the pond can cause some of the released ammonia and carbon dioxide to be lost to the atmosphere through volatilization.

Second, stocking a herbivorous fish (e.g., Tilapia rendalli or silver carp, Hypophthalmichthys molitrix) with a filter-feeding fish such as Nile tilapia can provide plant N and P more quickly than would otherwise be available from passive leaching and slower microbial decomposition.

Plant material, such as napier grass (Penniselum pururеum), added to the pond is fed upon directly by the herbivorous species, accelerating the decomposition process and release of algal nutrients through fish excretions. Rapid plant consumption would reduce both the oxygen demand in the pond from the added organic matter, and the amount of color-forming soluble organics released during decomposition.

Environmental Impacts

Adding green manures to a pond has several direct consequences for pond ecology. First, the amount of light energy in the water column is diminished through absorption both by the dark water color imparted by dissolved organics and by the suspended particulate matter from decomposing plants. Second, there is an additional microbial oxygen demand to decompose organic matter introduced into the pond.

The increased need of DO for microbial respiration is often made more acute by reduced oxygen production from algal photosynthesis because of decreased light availability. The severity of these impacts on pond ecology depends in part on the method used to put the green manure into the pond. Direct application of chopped plant biomass is likely the least effective way to fertilise a pond, except when culture organisms consume the plants directly.

Otherw ise, the chopped plants can exert a high oxygen demand, increase water color, and physically block out sunlight for algae. Partially decomposed plant material eventually settles to the pond sediments, promoting anoxic conditions in bottom waters while filling the pond with organic matter. Extensive use of green manures could necessitate mechanical aerators or other means of pond reaeration.

Negative ecological consequences of using green manures can be reduced by decomposing the plant matter in bins outside the pond and devising an efficient collection system for the leachate. In this way, atmospheric oxygen is used for decomposition and the partially decomposed particulate matter (which could be used to improve agriculture soils) is kept out of the pond.

Furthermore, if the leachate can be exposed to sunlight for several days before it is used to fertilise the pond, the breakdown of soluble organics will be faster and there will be less color imparted to the pond water . Although some CO_2 and ammonia is lost to the atmosphere, minimizing negative environmental impacts makes leachates more effective fertilisers.

Farm Costs

The deciding factor on the utility of green manures is probably based on farm-specific economics. There are several costs and labor demands which must be considered before adopting a green manure fertilisation strategy. For example, in addition to construction and maintenance costs for compost bins, the potential requirement for mechanical aerators should also be included in the budget analysis.

Although in-pond composting might be less expensive to set up, the associated environmental impacts discourage that approach. The plants used as green manures also come at a price. There are both out-of-pocket and labor costs incurred for growing, maintaining, harvesting, and transporting plants to composting areas.

These expenses can be reduced by growing marketable food crops (e.g., vegetables) along the pond banks, and then composting the nonconsumable parts of plants otherwise raised for food. There are essentially no additional production costs of this approach, and it is a convenient way to recycle nutrients from decomposing plants back into living biomass (e.g., algae and other natural foods).

It is better to compost fresh, green plant matter, because dried by-products (e.g., rice straw) have already lost most of their leachable nutrients . As an alternative to composting, some agricultural by-products may be suitable as a supplementary food source for farm animals. A portion of the P, N, and C contained in the plant material is incorporated in animal biomass, while another portion is excreted, which then becomes available for pond fertilisation.

For all these different ways of integrating plant crops, livestock, and aquacultural production systems, there are labor costs involved. If the farmer has to hire the labor, then these expenses become quantifiable and more readily incorporated into a comprehensive economic analysis. But if, for example, the farmer plants legumes to be used as a pond fertiliser, then determinations of the actual price of available P, N, and C for algal productivity must include:

— all legume production costs (including seeds and/or transplanted cuttings);

— more profitable alternative uses for the land and water used to grow the plants; and

— more profitable uses of the farmer's time and labor (i.e., lost opportunity costs).

The appreciation of lost opportunity costs is of paramount importance for the farmer's economic development. A farmer who spends hours a day collecting and/or processing manures (both animal and green) for aquaculture should do so only if it makes economic sense to that farmer. Otherwise, it becomes a very high price to pay for algal nutrients if the land, water, and labor could be used for more profitable endeavors.

Chemical Fertilisers

Chemical fertilisers are concentrated sources of P, N, and C which are manufactured, distributed, and sold in package form. The ones used in aquaculture are identical to those used in terrestrial agriculture. The more commonly used chemical fertilisers for P include mono superphosphate (MSP), triple superphosphate (TSP), and the various combinations of nitrogen-phosphorus-potassium (N-P-K) fertilisers.

Note that N-P-K fertilisers are designated by the content weight of N, P_2O_5, and K_2O, so a 20-16-8 grade fertiliser has actual N-P-K contents of 20%, 7%, and 6.6%, respectively. In addition to the N-P-K fertilisers, other N fertilisers include urea and sodium nitrate.

Fertilisation to increase inorganic carbon availability is accomplished by adding carbonaceous minerals such as lime ($CaCO_3$) and sodium bicarbonate ($NaHCO_3$). Lime should be added with the appreciation that adding too much calcium facilitates the precipitation of $CaCO_3$ during afternoon increases in pH.

Nutrient Availability

Since under typical pond water conditions these fertilisers eventually release 100% of each nutrient, the amount of soluble P, N, and C made available for algal uptake is readily calculable. It is better to dissolve the fertiliser in a bucket or put granules on a submerged platform to maximize soluble nutrient availability in the water column.

It is important to remember that only part of the chemical fertiliser actually contains the desired algal nutrient(s). For example, urea is 46.7% N, and therefore 1 kg urea contains 0.467 kg of urea-N. Due to the prevalence of counterfeit fertilisers in some developing countries (e.g., some bags of urea produced in Thailand were found to contain considerably less than 46.7% N), it may be prudent to verify either the source or the actual nutrient concentration when first using a particular chemical fertiliser.

Environmental Impacts

The dominant impact on pond ecology from the use of chemical fertilisers is the corresponding rise in algal productivity, the exact result a farmer wants from pond fertilisation. Since no additional organic matter is added to the pond, deoxygenation is not a problem. In highly productive, warmwater ponds (afternoon DO levels around 20 to 30 mg l^{-1}) fertilised only with urea and TSP, predawn DO concentrations typically average around 3 mg l^{-1} during a five-month grow-out of Nile tilapia.

The pond microbial community has sufficient DO to metabolize the daily production of algae-derived detritus and soluble organic molecules. In addition to not containing organic matter, a second benefit of chemical fertilisers over most organic fertilisers is that they do not affect light availability to algae. Chemical fertilisers dissolve in water, so there is no additional particulate matter suspended in the water column, such as that observed with manures.

Nor are there tannins or other dissolved organic substances which impart significant color to the water, as is the case with green manures and animal manures from ruminants. Furthermore, the rate of P adsorption by pond sediments can be reduced more quickly by applying a P fertiliser such as TSP. Organic matter on the pond bottom adsorbs P less readily than bare inorganic sediments.

Adding TSP will more rapidly fill remaining P adsorption sites in the sediments. Reducing net P losses to the pond sediments will increase the availability of subsequent P fertilisation to the pond's algal community.

One common misconception is that chemical fertilisers increase acidity in outdoor culture ponds. In fact, the addition of nitrates, phosphates, and carbonates actually increases alkalinity in water. A likely source of this confusion may be the misapplication of a laboratory investigation by Hunt and Boyd which examined the decomposition of urea and ammonia-based fertilisers in water.

Ammonia-based fertilisers and urea both add ammonia to water. Urea can also be directly utilized by many species of algae, bacteria, and fungi , or broken down through enzymatic reactions using urease. Equation 2 gives the general chemical reaction for the production of ammonia from urea:

$$CO(NH_2)2 + H_2O \rightleftarrows CO_2 + 2NH_3 \quad (2)$$

urea urease

Ammonia in water goes between the un-ionized (NH_3) and (NH_4^+) ionized forms, Equation 3 shown:

$$NH_3 + H_2O \rightarrow NH_4^+ + OH \quad (3)$$

This reaction shows that adding ammonia to water actually increases alkalinity with the addition of each hydroxyl ion (OH^-,). If ammonia remains in oxygenated water, however, the slow microbial oxidation of NH_4^+ to nitrate (NO_3^-) releases two hydrogen ions (H^+) for every molecule of ammonia, thus causing a net increase in the water's acidity.

If ammonia is taken up biologically before it gets slowly oxidized to nitrate, then H^+ ions are not released and there is no additional acidity. For example, in the presence of sunlight, ammonia released from urea (or from any other source of ammonia, including the decomposition of manures and detritus) is rapidly taken up by algae during photosynthesis. Alkalinity increases with the addition of OH- when ammonia chemically reacts with water (Equation 3), but algal uptake of NH_4^+ prevents any further oxidation to nitrate, and therefore H^+ ions are not produced.

Recent research confirmed both the loss of alkalinity when urea is decomposed in the dark, and the expected increase in alkalinity when sunlight is present (Figure 4). Vlek and Craswell also reported a similar increase in pH and alkalinity in flooded riçe fields fertilised with urea. Because these increases in alkalinity are due to additions of OH^-, there is no effect on total DIC beyond the additional CO_2 which is produced during urea decomposition (Equation 2).

Because Hunt and Boyd's study was conducted in the dark, its practical application to pond fertilisation is severely limited. Since algae utilize ammonia, regardless of the source, research clearly shows that acidification does not take place in fertilised ponds. In fact, if ammonia were a significant source of acidity in culture ponds, the continual release of ammonia from decomposition should have an observable effect.

Decades of fertiliser research has yet to document acidification in a pond which could be attributed directly to ammonia oxidation. Figure 4 indicates that it could take weeks for acidification to occur even in the dark. As a practical issue, the need to replace ammonia-based fertilisers with sodium nitrate to prevent pondwater acidification has been overstated. Conditions are rare which would permit acidification of pond water due to fertilisation with ammonia.

The accumulation of ammonia in unproductive ponds (e.g., due to inorganic turbidity) which have been heavily overfertilised with ammonia-based fertilisers (including

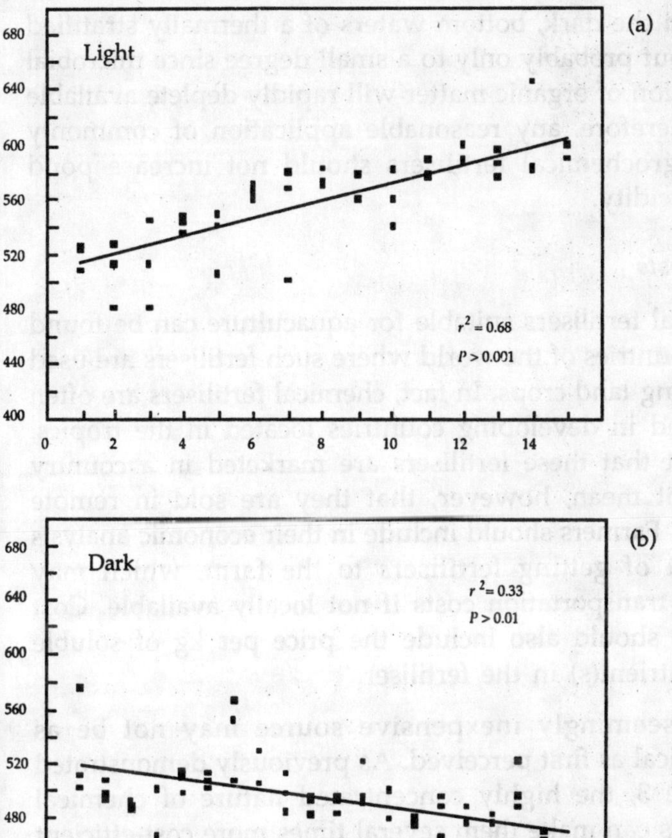

Figure 4. Changes in total alkalinity with the addition of urea, an ammonia-based chemical fertilizer. Alkalinity increases in the presence of light (a) because of photosynthetic uptake of ammonia and decreases in the dark (b) because of the oxidation of ammonia to nitrate and nitrite.

manures) can cause both a loss of alkalinity through the oxidation of ammonia to nitrate and potentially toxic levels of un-ionized ammonia. Oxidation of ammonia can also occur in the dark, bottom waters of a thermally stratified pond, but probably only to a small degree since microbial respiration of organic matter will rapidly deplete available DO. Therefore, any reasonable application of commonly used agrochemical fertilisers should not increase pond water acidity.

Farm Costs

Chemical fertilisers suitable for aquaculture can be found in all countries of the world where such fertilisers are used for raising land crops. In fact, chemical fertilisers are often produced in developing countries located in the tropics. The fact that these fertilisers are marketed in a country does not mean, however, that they are sold in remote villages. Farmers should include in their economic analysis the cost of getting fertilisers to the farm, which may include transportation costs if not locally available. Cost analysis should also include the price per kg of soluble algal nutrient(s) in the fertiliser.

A seemingly inexpensive source may not be as economical as first perceived. As previously demonstrated in Table 3, the highly concentrated nature of chemical fertilisers can make them several times more cost-efficient sources of algal nutrients than purchased manures, even though manures are considerably less expensive on a per-kg basis. Anderson also reported the economic benefit of chemical fertilisers over organic sources.

However, purchasing the least expensive source of soluble N and P may be difficult in some cases. The market cost per kg of fertiliser generally increases as bag size decreases, and poorer farmers may not have the financial resources to purchase the more economical 50-kg bags.

Furthermore, N-P-K fertilisers, typically available in remote villages, tend to be more expensive sources of soluble N and P than urea and TSP, respectively. This is in part due to the inclusion of K in the fertiliser, which has never been shown to limit algal productivity and represents an unnecessary expense to the farmer.

If individual farmers cannot afford to purchase fertilisers in bulk, then joining a group of farmers or a farming cooperative would allow them to purchase only what they need. With livestock-land crop-aquaculture integrated farming systems, it is more common to put chemical fertilisers on the land crops and manures in the ponds. Ecologically, and perhaps economically, it makes more sense to do the reverse.

Manures decompose more readily in air, which has about 20 times more oxygen than does water at saturation. And rather than filling up ponds with organic matter and depleting waters of oxygen, composted manures applied on land can improve soil quality as well as provide a source of plant nutrients. Chemical fertilisers could supplement manure treated soils as required.

On the other hand, chemical fertilisers applied to land crops can easily wash out once in solution; the high N and P content of agricultural drainage water is evidence of these losses. Directing this drainage into a culture pond would help reduce this potential loss to the farm while helping protect downstream environments from unwanted eutrophication.

From a whole-farm perspective, adding chemical fertilisers to the pond (to intentionally create eutrophication) and using pond water to irrigate land crops should further increase nutrient utilization efficiencies. A farm economic analysis which includes an assessment of ecological impacts should indicate which fertilisation strategy is the most cost-effective.

Other economic benefits of using chemical fertilisers for pond fertilisation include their consistency and predictability in nutrient concentrations, their long shelf-life when properly stored, and their relative ease of application. Labor costs involved with applying chemical fertilisers are lower than those for organic fertilisers. Manures tend to be noxious, heavy, bulky, and required in much greater amounts. For example, a typical fertilisation rate with chicken manure is about 500 kg ha^{-1} wk^{-1}, which provides approximately the same amount of soluble N and P as 5 kg ha^{-1} wk^{-1} of urea and TSP, respectively.

However, the farmer's decision whether to use organic or chemical fertilisers should be based primarily on farm economics, which necessarily includes relevant aspects of pond ecology. Although algae do not question the sources of P, N, and C, the farmer should question the environmental costs and economic benefits of each potential fertiliser. The farmer should also consider several pond-specific characteristics which can affect the efficacy of a particular fertilisation strategy.

10

Promise of Biotechnology

Biotechnology means a set of revolutionary techniques. It has been the subject of public policy aspirations for the last two decades. Promises to make a significant contribution in enabling the development of, for example, better health care, enhanced food security through sustainable agricultural practices, improved supplies of potable water, more efficient industrial development processes for transforming raw materials, support for sustainable methods of aforestation and reforestation, and detoxification of hazardous wastes.

Biotechnology-related applications and products have penetrated all sectors of the economy. The technology has begun to overcome the bottlenecks that, in the last century, favoured chemical substitutes against biological ones. As the knowledge base of biotechnology consolidates, the number of platforms that will depend on it will multiply to generate new fields. Despite these developments, biotechnology does not seem to have taken root in many developing countries and the goals have not been attained.

Food insecurity, disease and poverty still ravage a huge section of the human population, mainly in developing countries. Public attitudes and political will towards biotechnology have changed in some regions of the world. Similarly, the wider applications of biotechnology are buried in the debate surrounding genetically engineered crops. It is also possible that biotechnology may be gaining wide use in other fields that are not part of the current debate.

Opportunities in Industrial and Environmental Biotechnology

Industrial and environmental biotechnology is a broad category of technologies that employ enzymes and microbes in a wide range of industrial and pollution control processes. Industrial biotechnology products and processes are likely to become as ubiquitous as those of the chemical industry today.

Some analysts compare the current status of biotechnology to that of chemistry in the 1870s when it had a limited range of applications. Today, industrial chemistry is found in the food processing, pharmaceutical, fuel production, textile, fertilizer, water and paper industries, among others. Industrial biotechnology is likely to develop the same way.

Enzyme Technology

Enzyme technology is going to play a crucial role in industrial biotechnology. This includes native and genetically enhanced enzymes likely to function in environments previously thought to be hostile, as well as the engineering of new metabolic pathways in organisms to empower them to play a new role.

For instance, enzymes have been developed for use in detergents and production of biofuels, vitamins, amino acids and fine chemicals. Novozyme has developed enzymes for use in animal feed, food, textiles, leather, oil/fat and meat processing, among others. It has over 700 products and 100 different types of enzymes and microbes, replacing chemical products that pollute the environment.

Similarly, Genencor is developing enzymes with improved performance in detergents and vitamin C, biofuel, sugar and biopolymer production, while Prodigene is manufacturing TrypZen, a recombinant trypsin used in wound care and food processing. Such innovations will cut the cost of production, the number of processing steps and energy spent. They are also likely to reduce the cost of investment, environmental pollution and demand for high-grade feedstock.

The antibiotic cephalexin synthesis is also involved a multistep chemical process but is currently reduced to a mild biotransformation. The biotechnological process uses less energy and input chemicals, is water-based and generates less waste. Some of these enzymes come from organisms that live in hostile environments, organisms generally referred to as extremophiles, such as those found in hotsprings, salty waters and polluted surroundings among others.

The organisms survive in these environments because they possess unique enzymes that support live-saving pathways, whereas in such environments most organisms would be killed. The enzymes could be harnessed for industrial use, such as detergents used in detergents, textile industry, pharmaceuticals and bioremediation processes. Firms such as Applied Molecular Evolution, Genencor and Maxygen are interested in extremophiles for their peculiar metabolism and evolution.

Bioplastics

There is significant interest in the production of plastics made from renewable resources because they are biodegradable and thus environmentally friendly. Of the 40 billion tons of global production of plastics, bioplastics accounts for only 500 million tons (roughly 1.25%). If their production costs could be halved, the amount of bioplastics produced in 2010 could be trebled. The main platforms for bioplastic production include the use of microbes, plants and animals to produce desired plastic polymers, and the use of microbes and/or enzymes to convert carbohydrates and/or proteins into desired plastics.

For example, Cargill-Dow Chemical Company employs enzymes to produce Ingeo or Nature Works PLA, a polylactic acid (PLA) product made from glucose. It has commissioned a $300- million plant that can manufacture 140,000 tons of Ingeo, for use mainly in packaging. Similarly, DuPont employs a transgenic bacterium containing biochemical pathways from three different micro-organisms to convert (maize) glucose syrup to 1,3-propandiol, used to manufacture a polyester called Sorona, a copolymer, made from 1, 3-propandiol and terephthalate (oil product).

Some bacteria synthesise and accumulated polyhydroxyalkanoate (PHA), used to make bioplastic, up to 80% of their weight. The firm Metabolix is developing plastics from PHA and has genetically engineered plants to produce PHA. Metabolix planned to start commercial production of PHA by the end of 2003. There is increasing interest in bioplastic production as it is biodegradable and comes from renewable sources. However, the cost of the final products remains higher than equivalents made from fossil fuels.

Biofuels

Although ethanol powered the first car of Henry Ford, very few cars today use ethanol (alcohol). Brazil uses fuel blends with up to 20% of ethanol, while in the United States nearly a tenth of all motor vehicle fuel sold is blended with up to 10% ethanol. Ethanol is produced from cane sugar in Brazil and from maize in the United States. The US ethanol production is expected to reach 75 billion litres a year by 2020 from the current 9 billion.

In January 2003, Iogen, a Canadian firm, opened a pilot plant that converts straw into ethanol using cellulase. Iogen's main partners and investors in the EcoEthanol project are Shell, Petro-Canada and the Government of Canada. Canada intends to quadruplicate its ethanol production, up to 1 billion litres, between 2000 and 2005. While the production, transport and consumption of gasoline generate 11.8 kg of carbon dioxide per gallon (3.8 litres), ethanol generates 7 to 10 kg of carbon dioxide if conventionally produced, and only 0.06 kg if one relies on bioprocesses.

Bioremediation

Bioremediation refers to techniques that employ living organisms, such as microbes and plants, to extract, eliminate and/or bind toxins in forms that are not harmful to the environment. These include biostimulation, biotransformation, biostabilisation and biofiltration. For instance, microalgae are used in ponds to eliminate nitrogen and phosphorous, and aquatic plants (e.g. water lentils) are used to extract heavy metals in industrial effluents.

These natural processes have been employed for many years to eliminate pollutants. Modern biotechnology techniques promise to enha nce the performance of these natural processes in pollution control. For example, mercury is a highly toxic metal that accumulates in the food chain when released in water, for example in the Minamata accident, where inhabitants of the Japanese island of Kyushu suffered the toxic effects of fish poisoned by mercury-rich industrial effluents.

Since naturally thriving mercury-tolerant bacteria are rare and cannot be grown easily in culture, researchers at Cornell University inserted the metallothionein gene into Escherichia coli, which grows well in culture. The genetically engineered bacteria are placed inside a bioreactor that efficiently removes mercury from water.

The bacteria are later incinerated and the accumulated mercury is recovered. Existing techniques of mercury removal are expensive and inefficient. Phytoremediation refers to the use of plants to remove pollutants from water and soils. There are about 1.4 million polluted sites in Western Europe alone. Current techniques are costly and destroy soil structure. The use of plants that can store 10 to 500 more pollutants in their leaves and stems is cheaper and stabilises the soil structure.

Above all, the metals can be recovered from ashes and reused. There are many hyperaccumulating plants. These plants accumulate lead, zinc, nickel, copper and cobalt, among others, at levels toxic to other plants. For example, Sebertia acuminata can contain up to 20% of nickel in its sap (nickel is generally toxic to plants at a concentration of 0.005%). Similarly, the fern Pteris vittata accumulates arsenium while conserving a very rapid growth and a high biomass.

The firm Edenspace has acquired the commercialisation rights of the fern (now called edenfern™) for use in phytoremediation. The potential market for Phytoremediation in the United States alone was estimated at $100 million in 2002. Current research seeks to enhance, through genetic engineering, the ability of plants to accumulate heavy metals.

For example, the introduction of two Escherichia coli genes encoding an enzyme involved in transformation of arsenate into arsenite into the genome of Arabidopsis thaliana enables the plant to accumulate three to four times more arsenium than the normal plants. Similarly, tobacco plants with bacterial genes that control the synthesis of an enzyme-detoxifying trinitrotoluene, TNT, have been developed. Such initiatives could improve bioremediation processes through the introduction of genes controlling specific degradation pathways.

Biofertilisers and Biopesticides

Nitrogen supply is a key limiting ingredient in crop production in many African countries. It is often not available and/or beyond the reach of many poor farmers, especially those in rural areas. However, biological nitrogen fixation (BNF), the fixing of atmospheric nitrogen by microbes and making it available to plants, could be harnessed to improve the soil fertility and productivity of crops. These microorganisms are often referred to as biofertilizers.

However, biofertilizers also include microorganisms that solubilise phosphorus to make it available for plants. Many microorganisms have the ability to fix nitrogen. These include Azospirillum, Azotobacter, Rhizobium, Sesbania, algae and Mycorrhizae, while P. striata, and B. megaterium and Aspergillus are among other microorganisms that solubilise phosphorus. In return, the plant provides these organisms with a favourable habitat and a carbon source in a symbiotic relationship.

It is this relationship that is critical in seeking to broaden the use of biofertilizers in association with many food crops. Biofertilizers have been used in Kenya, the United Republic of Tanzania, Zambia and Zimbabwe. They are easily produced locally and the technology needed to produce them is not complex. In some countries, the demand has often outstripped production of the pilot plants.

Expansion of these pilot plants could help improve food productivity in Africa. The use of biopesticides in the control of pests is well established. For example, sterile tsetse flies (the vector of sleeping sickness) were used to control and eliminate the tsetse fly population on the island of Zanzibar. Similarly, the cassava mealybug, Phenacoccus manihoti, was effectively controlled using a wasp, Apoanagyrus lopezi, from Latin America, and this work was awarded the World Food Prize.

The bacteria Bacillus thuringiensis (Bt) has been used by farmers to control worms and insects for many years. Nematodes, bacteria, fungi and viruses may be used to control industrial, home and farm pests. On a large scale, the use of biopesticides has remained small, representing only a small fraction of the global $8 billion pesticide market. Bacillus thuringiensis (Bt) alone accounts for 90% of the $160 million biopesticides market.

The biopesticide market is driven by consumer, retail and government demands for reduction in use of chemical fertilizer use. The limiting factors include lack of spectrum (few targets), slow killing rate, batch variations, high sensitivity (to soil types, chemicals, temperature and moisture content) and low stability (short shelflife and high storage needs).

Biofertilizers and biopesticides present developing countries with an excellent opportunity to enhance their crop yields. Countries such as Bangladesh, Brazil, Kenya, the United Republic of Tanzania, Zimbabwe and Zambia have had successful pilot plants for the production of biofertilizers, and demand has often exceeded production.

Biopesticides, too, could help increase crop yield, reduce import bills and increase export earnings. Taken together, they could provide an affordable source of agricultural inputs, especially in rural areas where chemical inputs are unavailable or resource-poor farmers cannot afford them.

Other Industries

Other industries such as mining are already benefiting

from biotechnology. Bioleaching is a common technology in developing countries' mines. The small mining sector, often targeting small mineral deposits, could use bioleaching technology to improve the quality of the final products and reduce waste associated with mechanical cracking. In other cases, amethyst, agate, diamond and gold mining still use harmful chemicals.

Finding biotechnological solutions will increase the value and earnings from this sector, as well as reduce environmental degradation. The leather and textiles industries have been among the major environmental polluting industries. The use of enzymes will reduce industrial discharge through recycling of water, cut down the electrical and water bills and improve the quality of the final products.

Plants need not be rebuilt, but simple adjustments and replacement of harsh chemicals with biological systems are sufficient. With minor additions, enzymes and microbes could easily be produced locally. With a reduced clean-up bill, increased earnings and turnover, the industry will be set to become competitive. Paper production plants in some developing countries have either been closed or are uncompetitive. However, biotechnology presents this sector with many advantages that were never available before. The use of microbes and enzymes could replace chemicals, resulting in water and heat savings and improved quality of paper.

Genetic engineering may produce designer wood that will grow faster and, when processed, require few steps, resulting in extra savings and improved quality of paper. Many of the paper manufacturing plants that are currently uncompetitive could soon become exporters of paper. The most promising areas for many developing countries will lie in approaches that add more value to their raw materials. For example, technologies that will convert cassava into export products (e.g. plastics, sweeteners or fibres) will empower many poor farmers who currently do not have an international market for their products.

These fibres or polymers will be used to generate bags, plates and other utensils that have a higher value than the raw materials. Biotechnology could present a means by which to indirectly market products that are currently difficult to sell. With a market for tubers, their production could exceed that of cereals in no time, in many developing countries. Another promising application for developing countries lies in the conversion of waste into useful products.

Specifically, food waste may be broken down into amino acids, fuels and fertilizers that would benefit the rural and urban poor. Unlike the pharmaceutical industry, many developing countries could easily enter this market. The use of microbes and enzymes will be key in this revolution, and developing countries need to seize the opportunities. Many developing countries aspire to industrialisation.

The desire to migrate from raw material exports to processed products and increase their share of international trade provides the driving force. The ability to export finished products is partly dependent on technologies used, which may determine the quality of the products. Trade incentives such as the US Africa Growth and Opportunity Act of April 2000 should provide a basis for facilitating the transition towards industrial processing.

Asia and Latin America have fast-growing and robust textile industries, partly because there is a ready market for the products. Biotechnology promised to cut the costs of investments while improving the quality of products. It also promised to provide flexible processing and manufacturing platforms that could easily be modified and adapted.

In addition, it was going to reduce industrial stockfeed quality demands, waste production, energy, water and use of hazardous materials in production among others. In addition, the demand to meet market needs should force these countries, some with emerging and others with already growing markets, to seek alternative technologies. Industrial biotechnology is likely to play an important role.

Agricultural and Food Biotechnology

Plant biotechnology includes microbiology (e.g. biopesticides and biofertilizers), tissue culture (e.g. the clonal multiplication and production of planting material), marker-assisted breeding and disease identification and genetic engineering (e.g. the transfer of genes for one organism to the other, and deactivating or activating gene expression).

The in vitro (in the flask) growth of plant tissues and/or organs, followed by the clonal multiplication of the relevant plants to supply genetically identical and disease-free planting materials (referred to as tissue culture) to farmers, horticulturists, forest-tree growers and nurseries, is by far the most widely adopted biotechnology in developing countries.

For example, tissue culture has been used in oil-palm in Malaysia, Côte d'Ivoire and Indonesia, banana and plantain in Central Africa and Latin America, tubers and root crops (e.g. cassava in sub-Saharan Africa), tobacco, legumes, fruit trees, and many flower species and varieties in Colombia, and roses in Ecuador and Zambia. Tissue culture could be enhanced by advanced biotechnology techniques.

All the organisms used in agriculture and animal husbandry nowadays are the products of genetic modification for over 10,000 years through domestication, selection and artificial breeding methods aimed at generating plants and animals with improved performance. Tools such as artificial hybridization and forced mutagenesis have led to several crop plant varieties and animal breeds that may not have existed in nature. For example, the current commercial wheat is a product of at least 11 different varieties.

However, exchange of genes between biologically unrelated organisms does not occur in nature. Genetic engineering techniques permit selected genes to be moved between two or more organisms, even across different kingdoms (e.g. plants and animals), to generate transgenics or genetically engineered organisms.

Some examples of genetic engineering organisms are the incorporation of genes for insecticidal proteins of Bacillus thuringiensis (Bt) into the genomes of several crop species such as tobacco, maize, potato, rice and cotton, among others; genes for the synthesis of beta-carotene from daffodil and a bacterium into rice; genes for human haemoglobin into tobacco plants; genes for human milk proteins into rice; and gene for the hepatitis B surface antigen (HbsAg) into a yeast (Pichia).

Few developing countries are able to carry out the whole spectrum of research and development activities leading to the commercialisation of genetically engineered organisms. Many countries have limited human resources, institutional capacity, and legal and regulatory regimes to actively pursue research in agricultural biotechnology. However, the economies of many developing countries still largely depend on agriculture (for food supply, export and employment) and do not have or have very small subsidies from government. Their farmers are facing the following challenges to:

— Increase their production productivity and competitiveness at national, regional and international levels (within the framework of fair trade regulations);

— Protect environment and biological diversity, while reducing agricultural inputs (water, fertilizers and biocides), improve soil fertility and conservation (e.g. biological nitrogen fixation), and increase nitrogen and phosphorus absorption by crops;

— Diversify agri- food production so as to meet the changing needs of the consumers and food industry.

Advanced agricultural biotechnology can contribute to meeting these challenges. Current genetic modification (engineering) has provided protection against herbicides (for easy management of weeds) and some pests or both herbicides and pests.

Work is ongoing to protect plants against devastating insect, viruses, bacterial and fungal diseases as well as drought, cold, and salt levels. In addition, food crops with improved nutritional or health values are being designed. Biotechnology has to help, especially resource poor farmers, to produce more food in a sustainable way.

The use of all technologies available to enhance food production may have a greater impact than one technology that may not suit different societies. Adoption of any of the biotechnology-related products, if successful, could enhance acceptance of other advanced biotechnology products. Unless producers/farmers see the benefits of the new technologies in terms of increased yields, the productivity of crops and animals and profitability adoption will be met with scepticism.

Health-related Biotechnology

This is a field that has received significant attention from industry, government and international bodies. It is also the area where biotechnology has made significant inroads both in developed and in developing countries. The field, with the exception of animal cloning and stem cell research, is generally less controversial. Developing countries, most of which are in the tropics, have a high disease burden. Some of the major health diseases include malaria and tuberculosis, which spread fast in countries with poor housing and sanitation, as well as HIV/AIDS.

The economic development of nations, where millions of able-bodied people spend a significant proportion of their time bedridden while thousands more die from preventable and treatable diseases, is likely to be compromised. There are few alternative drugs for some of the most devastating diseases (e.g. malaria) in poor countries as the economic cost of producing drugs may exceed the proceeds from their sales. Some of the biotechnology platforms promise to cut down the cost of drug development.

Currently, the rate at which disease-causing organisms are developing resistance to the few available drugs threatens to outpace drug development. Biotechnology tools such as genomics promise to increase the pace of target molecule validation, which will in turn shorten the time required to develop a number of drugs for the same disease.

Antibacterial Compounds

There are over 5,000 antibiotic substances known today, with global production in excess of 30,000 tons and a total market value of $24 billion. Bacterial resistance to antibiotics has increased mainly due to increased use/abuse and similarities in antibiotic properties and structure, which lead to crossresistance. This raises the cost of health care. In the United States alone, antibioticresistant bacteria result in about $4.5 billion extra health care expenses.

The search for completely different compounds that attack bacteria through new mechanisms such as fluoroquinolones, quinoprisitin, dalfoprisitn, linezolid, ketolides and glycylcyclines is of great interest. A number of compounds have been found in animals, for example anti- microbial peptides such as magnainin, isolated from frogs. There is also increased interest in the search for specific genes in the sequenced genomes of major pathogenic bacteria to identify gene(s) coding for key but unique metabolic processes in the pathogenic microorganism.

An inhibitor molecule can then be engineered to attack such a process. Experts estimate that bacterial genomics have already given about 500 to 1,000 new broad-spectrum antibacterial targets. In addition, bacteriophages are making a comeback and can be of significant help for a few specific applications.

Another promising approach has been the combining of current antibiotics with compounds, "guardian-angel", that neutralise antibiotic-resistant bacteria. Clavulanic acid is one such compound, although it is ineffective in protecting the cephalosporins, frequently used in hospitals. The beta- lactamase inhibitors already developed have not progressed to clinical use because of their high production cost.

Protein Engineering

The modification of the activity of proteins is another research area of interest. For instance, Genencor is designing tumour-destroying proteins as well as proteins that will prorect the immune system against viruses and cancers, just as vaccines do.

Similarly, Maxygen has produced effective versions of interferons alpha and gamma, yet to be tested in people, and is developing proteins that would behave as vaccines against bowel cancer and dengue fever. Others include Viracept, a protease inhibitor for HIV (by Agouron) and

Relenza™ (by Biota Holdings), an inhibitor of neuraminidase of the influenza virus.

Genome Analysis

There is a strong belief that biotechnology tools, such as genomics, will make a significant impact in the near future. With the human and some animal genomes almost sequenced and tools for analysis of sequences developed, the pace of target evaluation and product/service development is likely to increase.

It may soon be possible to treat inheritable diseases by correcting genetic defects (manipulation of the genetic information of the cells in the laboratory, with their subsequent sending back into the patient) and develop personalised treatment regimes. The developments in genomics hold a lot of value for health.

If genes that enhance resistance to environmental stress and diseases are identified, it will be possible to activate them. Similarly, it is possible, in some cases, to predict the likelihood of populations or individuals succumbing to known diseases, such as cancers, using genetic information. This could lead to personalised treatment or designing drugs for populations of known genetic make-up.

Food and Nutrition

Food and health have always been closely linked - and not only in popular belief. Both foodstuffs and medicines have the power to heal bodily dysfunctions, while an imbalance can disrupt our well-being. At the very foundation of human health and well-being lies the ecosystem of intestinal microorganisms. This complex microflora, comprising a wide range of different bacterial species, plays several roles: supplying the human host with additional value from foodstuffs; protecting against intestinal infections; and contributing to the development of the immune system.

Many health-improving properties of certain foodstuffs are already well known: dairy products may strengthen the immune system; fruits and vegetables contain vitamins that protect humans against infections; meat and fish deliver proteins important for the growth and development of the young body; fibre-rich foodstuffs are important for the intestinal transport of digested food; and plant hormones have a long-term protective function against cardiac diseases and, probably, cancer.

Nutraceutics

Nutraceutics are functional foodstuffs capable of modifying one or more organic functions favourably, in addition to their nutritional effect. In the 1990s, research on functional foods led to products that were found to be "anti-cholesterol" (oil derived from maize) and "anti-oxidant" (a grapevine synthesising more resveratrol) among others.

For example, David Sinclair and his team at Harvard University indicated in the journal Nature that resveratrol could lengthen the life of a yeast (Sacharomyces) cell by 80% by activating enzymes that prevent cancer, stave off cell death and boost cellular repair systems. This naturally occurring molecule builds up in undernourished animals and plants attacked by fungi.

The world market for nutraceutics was estimated to be about $14 billion in 1997 and with a growth rate of about 20% per annum. Nutraceutics are driven by consumer demand for healthier products. Therefore, food firms are using biotechnology techniques to help them capitalise on emerging demands by consumers concerned about the quality of food and its impact on health. This is important especially in rich countries where many consumers take a lot of vitamin pills and other food additives to correct the deficiencies of an unbalanced diet. Genetic engineering could be useful for producing food ingredients deprived of some undesirable elements or enriched with healthy substances to qualify as nutraceutics.

Probiotics and Prebiotics

Probiotics are microbial food additives or ingredients that restore a good balance of beneficial microbial flora in the gut. They consist mainly of lactic acid bacteria, bifidobacteria and yeasts. They have health-promoting impacts in the oral cavity, stomach, small and large intestine, and the vagina.

Prebiotics, unlike probiotics, are non-digestible carbohydrates such as fructo- and galacto-oligosaccharides that exert health-promoting effects by improving the characteristics of intestinal flora. Prebiotics, like dietary fibres, act as anti-constipation, faecal bulking and pH-reducing ingredients. However, their major mechanism lies in their support of probiotics.

The European Commission's cluster on Proeuhealth research is studying gastrointestinal-tract functionality and human health. The cluster aims to provide:

— A clearer understanding of the relationship between food and intestinal bacteria, and human health and disease;

— New molecular research tools for studying the composition and activity of intestinal microbiota;

— New therapeutic and prophylactic treatments for intestinal infections, chronic intestinal diseases and for healthy ageing;

— A molecular understanding of immune modulation by probiotic bacteria and examination of probiotics as vaccine-delivery vehicles;

— Process formulation technologies for enhanced probiotic stability and functionality;
— Commercial opportunities for the food and pharmaceutical industries.

For example, lactobacilli (bacteria) are known to affect immunomodulation and were reported to reduce the risk of STD infections, including HIV, in women. Increased understanding of the molecular factors of these organisms contributing health effects could allow the selection of probiotic strains, with enhanced protective or therapeutic effects.

Delivering the health benefits of probiotics and prebiotics to consumers depends essentially on their successful processing, viability, stability and functionality, as well as on storage. European researchers are exploring optimal process and formulation technologies for use in processing probiotics.

Biotechnology and Emerging Trends

Biopharming

Biopharming refers to the production of pharmacological products in genetically engineered plants. By 2003, about 300 trials of crops genetically engineered to produce various therapeutic products were initiated. These include modified tobacco plants that produce Interleukin-10 for the treatment of Crohn's disease, GM potatoes that produce antibodies for reducing the risk of rejection in kidney transplants, GM tobacco that produces vaccines against hepatitis B and drugs against HIV/AIDS, and potatoes that produce human insulin.

Other GM plant-produced substances include enkephalins, alpha- interferon, serum albumin and glucocerebrosidase. Clinical trials have begun on crop-grown drugs to treat cystic fibrosis, non-Hodgkin's lymphoma and hepatitis B. Various firms and research centres such as SemBioSys Genetics, Inc., Canada, Planet Biotechnology, Inc., United States and ProdiGene, Inc., United States, have demonstrated the effectiveness of plants as bioreactors.

For example, mice fed on potatoes expressing the beta-subunit of cholera toxin (CTB) were resistant to the cholera toxin. Given that CTB gives greater protection against cholera to humans than to mice, one expects higher levels of protection in human. Others include transgenic tomatoes containing a gene from Escherichia coli that can protect against diarrhoeal diseases.

Several laboratories around the world are working on their own versions of plantderived vaccines, using tomatoes, bananas and potatoes among other crops. Biopharming is mainly driven by a cost advantage. For example, medicinal products could be synthesised in plants at less than one tenth of the cost of conventionally manufactured drugs and vaccines. By the end of the current decade, biopharmaceuticals are projected to grow into a $20 billion industry.

It could ultimately bring down the cost of treating some diseases. Animals, too, have been engineered to produce materials for pharmaceutical and industrial use. Once manufacturing biopharmaceuticals in animals becomes efficient, the flow of more than a hundred protein-based drugs currently in advanced phases of clinical trials and many more that are in development in the laboratory would increase.

Food Taste Modification

The use of salt and sugar/sweeteners in food is generally intended to mask the poor taste of some foods. For example, grapefruit juice will not be sweet without added sugar and potato chips will not be flavourful without added salt. The search is on for molecules capable of tricking the taste buds on the tongue. For example, the compound adenosine 5′-monophosphate (AMP), which occurs naturally (e.g. in human breast milk), blocks acidic tastes. When AMP is added to certain foodstuffs, such as coffee and citrus juice, it prevents some of the acidic tastes from being felt by the tongue. Such activities could have applications in medicine manufacturing as well.

Several food firms, such as Coca-Cola Co., Kraft Foods and Solae, are interest in food flavour and have major deals with research firms involved in tricking the receptors on the tongue by accentuating or blocking certain elements in the food. This is useful since processed foods, such as canned soups, sauces and snacks, contain large amounts of salt to mask the bitter tastes, and soft drinks are sweetened to tone down the bitter taste of caffeine.

Health and nutrition concerns, fuelled by reports on obesity, diabetes, cardio-vascular diseases and hypertension, are driving the current interest in molecules that could trick the taste buds. Linguagen Corp has discovered about 20 compounds that block bitter tastes and has been granted patents to use four of the compounds as bitter blockers. Since humans have more than 30 different bitter taste receptors, finding a universal bitter blocker is nearly impossible.

Linguagen Corp is also trying to discover and market a natural sweetener to replace artificial ones such as aspartame or saccharine, which often leave a bitter aftertaste. The company plans to license bitter blockers in food, beverage and medicine manufacturers in the United States by early 2004. Other research firms, such as Senomyx, are also developing molecules that block bitterness and unpleasant smells and increase the salty taste.

Health and National Security

The development of "smart plants" (plants that can detect and/or remove toxins and pollutants) is very important, especially in improving environment, enhancing health and promoting national security. In combination with bioremediation uses, "smart plants" could detect heavy metals, microorganisms and other environmental changes. Such plants could be used in phytoremediation, while at the same time monitoring the changes in quantities or presence of toxins or pollutants.

For example, Aresao Biodetection, a biotechnology company in Demark, has genetically engineered Thale cress (Arabidopsis thaliana) for the detection of landmines and/or exp losives. The plant changes colour from green to red when it comes into contact with explosive materials. The company is also developing plants for detection of heavy metals. Biometrics, the automated methods of recognising a person on the basis of physical features such as face, fingerprints, hand, iris and voice, is already being deployed to enhance the identification and verification of individuals.

For example, Iridian Technologies's 'iris recognition technology' is already being used at some airports and borders for identification and verification of persons. It is more accurate, faster and easy to use than standard methods. The production of "Pocket", a portable, battery-operated protein analyser that may cost only $45 commercially. It could be used in immune assay experiments to detect several diseases, including HIV. Similarly, Lawrence Livermore National Laboratories announce the production of a portable DNA analyser for detection of infectious agents.

Many of such tools are likely to reach the market in the near future and reduce the costs of equipping, maintaining and running diagnostic laboratories in developing countries and improve health. For example, Integrated Nanotechnologies LLC has developed a DNA analyser to detect a host of infectious organisms, including anthrax and SARS. The analyser is self-contained and does not require operator adjustments between tests, thus making it easier to run. A lighter and commercial versio n is expected for release this year. Although designed for bioterrorism in developed countries, the analyser is a useful tool for developing countries too as it could cut down reagent needs and overcome current lack of trained personnel.

Scientist at Purdue University announced the development of a chip-sized version of a common detector used to identify proteins, DNA and other molecules. This could radically reduce the size of detection equipment, in a fashion similar to that in which the move from separate transistors to integrated circuitry changed to the size and power of computers and similar equipment. It is expected that in a decade or two, scientists will have invented tools that can analyse a human genome in a day and make it possible to profile entire populations.

For example, the Affymetrix GeneChip array may measure the activities of thousands of genes simultaneously. Cloning, especially of animals, may soon become a routine practice of some laboratories. For example, Aegen Bioscience, has developed a cloning chip that could denucleate, one of the first hurdles in cloning, more than a hundred cells at once.

Many other efforts are already taking place in various research centres. The Massachusetts Institute of Technology (MIT) has produced a 'biorubber' that could be used safely in engineering heart valves, blood vessels, livers and other elastic body tissues. A collaborative team has produced a 'biogel' that may change its characteristics in response to stimulus, a property that makes the biogel a candidate for construction of controlled drug delivery devices and similar tools. At the Department of Experimental Surgery at Berlin's Charité Hospital, scientists have developed a small bioreactor containing a matrix of hundreds of membranes, within they have coaxed human adult liver stem-cells to grow into complex living tissue remarkably like a healthy liver.

When the researchers feed a patient's blood through the bioreactor, the cultured liver cells take over all the normal, healthy functions of the patient's own diseased organ. The bioreactors are being use in clinics in Berlin and Barcelona to save the lives of patients whose own livers have stopped functioning but whose donor organs have not yet arrived. Others are targeting the liver's regenerative capacity in order to make many transplants unnecessary in the future by hooking patients up to the reactor so their own livers can take time off and recuperate.

Research is underway at the University of Pittsburgh to use the tissueculture techniques developed for the bioreactor to induce the body to grow new liver tissue on its own. The future trends in biotechnology are likely to be influenced by advances in the other technological fields. Some of the technologies that will make great impressions on biotechnology research and development may include information and communication technologies, materials technology, cognitive technology and nanotechnology.

The convergence of any of these or all of these and other fields with biotechnology is likely to spin off new technologies that will make a huge impact on industrial competitiveness and quality of life. Materials technology and nanotechnology are enabling the deve lopment of smarter, smaller, productive, efficient, durable, multi-purpose and user-friendly materials. Such materials will enable the construction of tiny devices that could be embedded in animal or plant tissue for controlled drug delivery, monitoring tissue activity, performance and location of the organism.

Genomics may enable individuals to predict the chances of catching a given disease, physical appearance and, perhaps, levels of intelligence. Similarly, the ability to produce replacement organs and better delivery and diagnostic systems will improve health. Taken together, these technologies will enable production of materials that will interact (identify, communicate, analyse and protect) with the user. Taken together, they will affect many industries.

Bibliography

Baxter, G.A., Areskoug, D., O'Connor, M., Ferguson, J., Crooks, S.R.H. and Elliott, C.T., "Direct immuno-biosensor analysis for streptomycin residues in honey and milk". *Poster Presentation at The Society for Food and Agricultural Immunology's Fifth International Conference: Agri-Food Antibodies '99*, Norwich, UK, September 1999.

Baxter, G.A., Elliott, C.T., Crooks, S.R.H. and O'Connor, M. "Biosensor analysis of sulphonamides". *Poster Presentation at Safe Beef. A seminar for the beef industry*, Loughry College, June 1997.

______________., Ferguson, J.P., O'Connor, M. and Elliott, C.T. "Detection of Streptomycin Residues in Whole Milk Using an Optical Immunobiosensor". *Journal of Agriculture and Food Chemistry*, 49:3204-3207, 2001.

______________., O'Connor, M., Haughey, S.A., Crooks, S.R.H. and Elliott, C.T. "Evaluation of an immunobiosensor for the on-site testing of veterinary drug residues at an abattoir". *Analyst*, 124(9):1315-1318, 1999.

Elliott, C.T., Baxter, G.A., Crooks, S.R.H. and McCaughey, W.J. "The development of a rapid immunobiosensor screening method for the detection of residues of sulphadiazine". *Food and Agricultural Immunology*, 11(1):19-27, 1999.

______________., Hewitt, S.A., Arts, C.J.M., Van Baak, M., Hellenas, K.E. and Johansson, A.M. "Use of biosensors for rapid drug residue analysis without sample deconjugation or clean-up: A possible way forward?" *The Analyst*, 123(12):2469-2473, 1998.

_______________., Crooks, S.R.H., Baxter, G.A., Haughey, S.A. and Situ, C., "Immunobiosensor assays using Biacore instrumentation: Combining method development, method evaluation and routine use". *The Society for Food and Agricultural Immunology's Fifth International Conference: Agri-Food Antibodies '99*, Norwich, UK, September 1999.

_______________., Traynor, I.M., Fodey, T.L. and Crooks, S.R.H. "Development of a high sensitivity, monoclonal antibody biosensor assay for multi B agonist detection in liver". *Poster Presentation at VI international conference on agri-food antibodies*, Prague, Czech Republic, October 2001.

Haughey, S.A., Baxter, G.A., Crooks, S.R.H. and Elliott, C.T. "Detection of ractopamine residues using the Biacore Q SPR biosensor". *Poster Presentation at The 116th AOAC International 2002*, Los Angeles,California,USA, September 2002.

_______________., Thompson, C.S., Traynor, I.M. and Elliott, C.T. "Use of the Biacore Q for the detection of ractopamine residues in porcine liver and urine samples". *Poster Presentation at VI International conference on agri-food antibodies 2001*, Prague, Czech Republic, October 2001.

Hellenas, K.E., Johansson, A.M., Elliott, C.T., Sternsjo, A. and Stenberg, E. "The Use Of Optical Biosensors For The Detection Of Veterinary Drug Residue In Foods-Progress Of The Foodsense Project". *Euroresidue IV*, 1:542-545, 2000.

Samsonova, J.V., Baxter, G.A., Crooks, S.R.H. and Elliott, C.T. "Biosensor Immunoassay of Ivermectin in Bovine Milk". *Journal of AOAC International*, 85(4):879-882, 2002.

Situ, C., Small, A.E., Fodey, T.L., Crooks, S.R.H. and McEvoy, J.D.G. "Development of Specific Antibodies to the Illegal and Unlicensed Antimicrobial Feed additives". *Poster Presentation at 4th Drug Residue Symposium*, Antwerp,Belguim, September 2002.

Index